AI and ML Techniques in Image Processing and Object Detection

Sandeep Chand Kumain · Maheep Singh ·
Lalit Kumar Awasthi · Raj Singh
Editors

AI and ML Techniques in Image Processing and Object Detection

Springer

Editors
Sandeep Chand Kumain
School of Computer Sciences
UPES Dehradun
Dehradun, Uttarakhand, India

Maheep Singh
Department of Computer Science
Doon University
Dehradun, Uttarakhand, India

Lalit Kumar Awasthi
National Institute of Technology Hamirpur
Hamirpur, Himachal Pradesh, India

Raj Singh
Department of Environmental Sciences
GITAM University
Visakhapatnam, Andhra Pradesh, India

ISBN 978-981-96-7444-2 ISBN 978-981-96-7445-9 (eBook)
https://doi.org/10.1007/978-981-96-7445-9

This Springer imprint is published by the registered company Springer Nature Singapore Pte Ltd.
The registered company address is: 152 Beach Road, #21-01/04 Gateway East, Singapore 189721, Singapore

This book is dedicated to the relentless seekers of knowledge in AI and ML, whose contributions continue to revolutionize image processing and object detection.

Preface

In the present digital world, the rapid advancement of modern techniques has significantly impacted the wide range of industries from medical imaging and agriculture to astronomy and video monitoring. These technological developments have reshaped how we perceive, analyze, and interpret visual data, enhancing automation, accuracy, and efficiency. This book brings together research contributions from experts across various fields to provide a comprehensive exploration of image analysis and object recognition methods. It guides readers through a diverse set of applications, highlighting their transformative role in healthcare, agriculture, security, space exploration, and urban planning. Each chapter addresses a specific topic, offering foundational concepts, technological progress, and real-world applications of image processing techniques.

The book covers the medical imaging applications, where deep learning models have demonstrated exceptional capabilities in automatic organ classification and cardiovascular disease prediction. These advancements in AI-driven diagnostics offer immense potential in assisting medical professionals with precise and efficient decision-making. The agriculture domain also benefits significantly from AI and ML integration. With intelligent data-driven approaches, farmers can optimize resources, predict crop yields, and improve sustainability. The fusion of AI with satellite imagery further enhances remote sensing applications, offering unprecedented insights into ecological monitoring crisis management and land-use analysis. Moving towards object detection, the book examines evolutionary techniques in computer vision, tracing the transition from classical algorithms to state-of-the-art deep learning frameworks. These methodologies are critical in areas such as autonomous vehicles, surveillance systems, and facial recognition.

Moreover, the book extends this exploration to astronomical image processing, where AI techniques have facilitated the identification of celestial bodies and cosmic phenomena, expanding our understanding of the universe. Furthermore, emerging topics such as generative models for image synthesis, disparity estimation in aerial datasets, and human action recognition in video surveillance demonstrate how AI continues to push boundaries in computational vision. The concluding chapters discuss speech processing for stuttering diagnosis and deep learning architectures for biomedical image analysis, showcasing the multidisciplinary applications.

This book aims to serve as a valuable resource for researchers, practitioners, and students keen on exploring the intersection of AI, ML, and image processing. The curated studies provide both theoretical insights and practical implementations, making it a useful guide for academicians and industry professionals alike. As we continue to witness rapid developments in artificial intelligence, we hope this compilation inspires future research and innovation, fostering advancements that will shape the next generation of intelligent systems.

Dehradun, India — Sandeep Chand Kumain
Dehradun, India — Maheep Singh
Hamirpur, India — Lalit Kumar Awasthi
Visakhapatnam, India — Raj Singh

Contents

Editors and Contributors

About the Editors

Dr. Sandeep Chand Kumain is an Assistant Professor at the School of Computer Sciences, UPES, Dehradun. He holds a Ph.D. and an M.Tech. in Computer Science and Engineering from the National Institute of Technology, Uttarakhand, India. His research interests include computer vision, image processing, and deep learning. Proficient in MATLAB, C, C++, Java, and Python, Dr. Kumain has authored over 15 publications in reputed international journals and conferences. e-mail: sandeep.kumain@ddn.upes.ac.in; sandeep.chandphd2021@nituk.ac.in

Dr. Maheep Singh is an Assistant Professor at the School of Computer Science, Doon University, Dehradun. Before joining Doon University, he served as an Assistant Professor at NIT Uttarakhand. He earned his Ph.D. from MNIT Jaipur. His research interests include salient object detection, image processing, machine learning, and security. With over 18 years of experience in teaching and research, Dr. Singh has published 30+ research papers in reputed international journals and conferences. e-mail: maheepsingh@nituk.ac.in; maheepsingh@doonuniversity.ac.in

Prof. Lalit Kumar Awasthi is the Vice Chancellor of Sardar Patel University, Mandi, Himachal Pradesh, India. Previously, he served as Director at NIT Uttarakhand. He earned his M.Tech. in Computer Science and Engineering from IIT Delhi (1993) and his Ph.D. from IIT Roorkee (2002). Professor Awasthi was the first faculty member of the Computer Science and Engineering Department at REC Hamirpur (now NIT Hamirpur) and contributed significantly for 25 years. He played a key role in establishing the B.Tech CSE program at NIT Hamirpur, developing laboratories, buildings, and the Computer Centre. He has also held various leadership roles, including Head of CSE, Director of the Computer Centre, and Dean (Students and Alumni). He was instrumental in the planning and establishment of Atal Bihari Vajpayee Government Institute of Engineering and Technology, Pragatinagar, overseeing infrastructure development, including workshops, hostels, and auditoriums. His research focuses on computer architecture, fault tolerance, parallel and distributed processing, checkpointing, mobile computing, ad hoc networks, and deep learning. Professor Awasthi has published over 250 research papers in reputed international journals and conferences, filed 14 patents, and secured three design registrations, with one patent successfully commercialized. e-mail: lalit@nith.ac.in

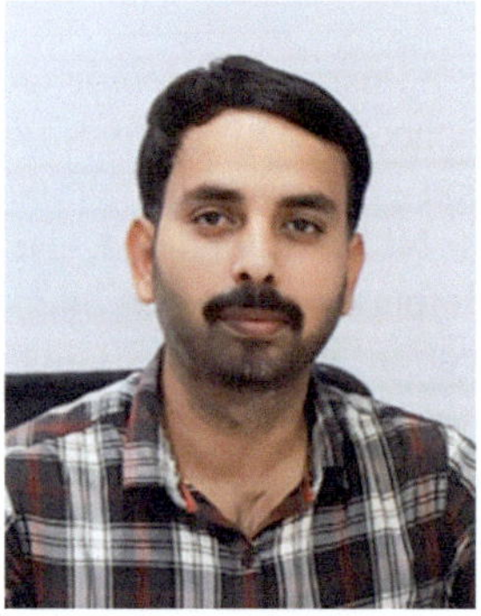

Mr. Raj Singh M.Sc., PGD, is a Research scholar in the Department of Environmental Science GITAM Deemed to be University, Visakhapatnam, Andhra Pradesh, India. He served as an Assistant Professor at Dr. K. N. Modi University, Newai, Rajasthan, and Tula's Institute, Dehradun, India. He is an alumnus of the Indian Space Research Organization (ISRO), Department of Space, Government of India, and also worked on a France-funded IDDRI project at BITS Pilani, KK Birla Campus, Goa, India. He is an active member of Wetland International, Wageningen, Netherlands, and the British Ecological Society London, UK. His expertise is in Space Science, Environmental Science, Wetland Ecology, Remote sensing, and GIS . He has published over 20 scientific articles in reputed national and international peer-reviewed journals, review, book chapters, presented six papers at international conferences, and edited three books with Jenny Stanford and Springer Nature publisher.

Contributors

Richa Baranwal National Institute of Technology, Delhi, India

Mehul Bhatia Dr. Vishwanath Karad MIT World Peace University, Pune, India

Yogesh Chandra Department of Physics, Government P.G. College Bazpur, Bazpur, Uttarakhand, India

Chhagan Charan ECE Department, National Institute of Technology, Kurukshetra, Haryana, India

Prajayshee Chauhan Dr. Vishwanath Karad MIT World Peace University, Pune, India

Deepak Dhillon School of Artificial Intelligence, Bennett University, Greater Noida, UP, India

Jihen Fourati Computer Sciences, ATES: Advanced Technologies on Environment and Smart City, National Engineering School, Sfax, Tunisia;
Computer Sciences, ATES: Advanced Technologies on Environment and Smart City, University of Gafsa, Gafsa, Tunisia;
National Engineering School of Sfax, University of Sfax, Sfax, Tunisia;
Physics, ATES: Advanced Technologies on Environment and Smart City, University of Sfax, Sfax, Tunisia

Sakshi Jaiswal Dr. Vishwanath Karad MIT World Peace University, Pune, India

Jyoti National Institute of Technology, Delhi, India

Balmukund Kanodia Department of Information Technology, Jadavpur University, Jadavpur University Second Campus, Kolkata, West Bengal, India

Dmitrii Kaplun Department of Automation and Control Processes, Saint Petersburg Electrotechnical University "LETI", Saint-Petersburg, Russian Federation

Monji Kherallah Computer Sciences, ATES: Advanced Technologies on Environment and Smart City, National Engineering School, Sfax, Tunisia;
Computer Sciences, ATES: Advanced Technologies on Environment and Smart City, University of Gafsa, Gafsa, Tunisia;
National Engineering School of Sfax, University of Sfax, Sfax, Tunisia;

Physics, ATES: Advanced Technologies on Environment and Smart City, University of Sfax, Sfax, Tunisia

Ekaterina Kopets Youth Research Institute, Saint-Petersburg Electrotechnical University "LETI", Saint Petersburg, Russian Federation

Indrajeet Kumar School of Engineering and Technology, Birla Global University, Bhubaneswar, Odisha, India

Sumit Kumar Department of IT, ABES Institute of Technology, Ghaziabad, India

Vinay Kumar Faculty of Agricultural Engineering, SKUAST-J, Jammu, Jammu and Kashmir, India

Vivek Kumar Department of CSE, THDC-IHET, New Tehri, Uttarakhand, India

Mudit Mittal Department of CSE, THDC-IHET, New Tehri, Uttarakhand, India

Ishan Narayan Academy of Scientific and Innovative Research, Ghaziabad, India;
CSIR—Central Scientific Instruments Organization, Chandigarh, India

Mohamed Othmani Computer Sciences, ATES: Advanced Technologies on Environment and Smart City, National Engineering School, Sfax, Tunisia;
Computer Sciences, ATES: Advanced Technologies on Environment and Smart City, University of Gafsa, Gafsa, Tunisia;
National Engineering School of Sfax, University of Sfax, Sfax, Tunisia;
Physics, ATES: Advanced Technologies on Environment and Smart City, University of Sfax, Sfax, Tunisia

Manjuleshwar Panda Delhi, India

Shashi Poddar Academy of Scientific and Innovative Research, Ghaziabad, India;
CSIR—Central Scientific Instruments Organization, Chandigarh, India

Rajat Rajoria Department of Information Technology, Jadavpur University, Jadavpur University Second Campus, Kolkata, West Bengal, India

Jyoti Ramola ECE Department, Graphic Era Hill University, Dehradun, India

Jyoti Rani GZSCCET, MRSPTU, Bathinda, India

Arjun Singh Rawat National Institute of Technology, Delhi, India

Debam Saha Department of Computer Science and Engineering, Calcutta Institute of Engineering and Management, Kolkata, West Bengal, India

Khawla Ben Salah Computer Sciences, ATES: Advanced Technologies on Environment and Smart City, National Engineering School, Sfax, Tunisia;
Computer Sciences, ATES: Advanced Technologies on Environment and Smart City, University of Gafsa, Gafsa, Tunisia;
National Engineering School of Sfax, University of Sfax, Sfax, Tunisia;

Physics, ATES: Advanced Technologies on Environment and Smart City, University of Sfax, Sfax, Tunisia

Partha Sarkar Department of IT, Sparsh Himalaya University, Dehradun, Uttarakhand, India

Sushil Sharma Faculty of Agricultural Engineering, SKUAST-J, Jammu, Jammu and Kashmir, India

Anu Singha Dr. Vishwanath Karad MIT World Peace University, Pune, India

Jaswinder Singh Punjabi University, Patiala, India

Pawan Kumar Singh Department of Information Technology, Jadavpur University, Jadavpur University Second Campus, Kolkata, West Bengal, India

Aishwary Varshney School of Engineering and Information Technology, Sanskriti University, Mathura, UP, India

Gaurav Verma ECE Department, National Institute of Technology, Kurukshetra, Haryana, India

Pankaj Verma ECE Department, National Institute of Technology, Kurukshetra, Haryana, India

Raj Kishor Verma Department of Computer Science and Engineering (Data Science), ABES Institute of Technology, Ghaziabad, India;
School of Computer Science and Engineering, Galgotias University, Greater Noida, India

Vaibhav Verma National Institute of Technology, Delhi, India

Jitendra Virmani CSIR—CSIO, Chandigarh, India

Alexander Voznesensky Department of Automation and Control Processes, Saint Petersburg Electrotechnical University "LETI", Saint-Petersburg, Russian Federation

Satya Prakash Yadav School of Engineering and Information Technology, Sanskriti University, Mathura, UP, India

Attention-Based Deep Neural Networks for Automatic Organ Classification from 2D CT Scan Images

Rajat Rajoria, Balmukund Kanodia, Debam Saha, Ekaterina Kopets, Alexander Voznesensky, Dmitrii Kaplun, and Pawan Kumar Singh

Abstract The area of medical imaging has seen a revolution in recent years due to the rapid advancement of deep learning (DL) techniques. Medical image analysis plays a key role in modern healthcare, helping in the accurate diagnosis and treatment of various conditions. Using deep learning, it is possible to solve complex medical image analysis problems with unparalleled accuracy and efficiency. In this research, we explore the potential of deep learning models for multi-class image classification of abdominal organ structures within 2D Computed Tomography (CT) images. We focus on the 2D views, namely axial, coronal, and sagittal, to facilitate lightweight model evaluation and deployment. Our paper presents a comprehensive analysis of the classification performance, with a particular emphasis on model generalizability, algorithm selection, and interpretability. In this paper, we investigate the classification

R. Rajoria · B. Kanodia · P. K. Singh (✉)
Department of Information Technology, Jadavpur University, Jadavpur University Second Campus, Kolkata, West Bengal, India
e-mail: pawansingh.ju@gmail.com

R. Rajoria
e-mail: rajatrajoria.ju@gmail.com

B. Kanodia
e-mail: kanodiabm@gmail.com

D. Saha
Department of Computer Science and Engineering, Calcutta Institute of Engineering and Management, Kolkata, West Bengal, India
e-mail: debamsaha.cse@gmail.com

E. Kopets
Youth Research Institute, Saint-Petersburg Electrotechnical University "LETI", Saint Petersburg, Russian Federation
e-mail: eekopets@etu.ru

A. Voznesensky · D. Kaplun
Department of Automation and Control Processes, Saint Petersburg Electrotechnical University "LETI", Saint-Petersburg, Russian Federation
e-mail: asvoznesenskiy@etu.ru

D. Kaplun
e-mail: dikaplun@etu.ru

S. Chand Kumain et al. (eds.), *AI and ML Techniques in Image Processing and Object Detection*, https://doi.org/10.1007/978-981-96-7445-9_1

of 11 distinct abdominal organs like the bladder, heart, kidneys, liver, etc., using state-of-the-art deep neural networks with an attention feature integrated. The results of this research contribute to the growing body of knowledge in medical image analysis, showcasing the potential of deep learning models in the context of multi-class organ classification. By offering improved accuracy, our findings may have implications for clinical practice, computer-aided diagnosis, and healthcare automation.

Keywords Biomedical image classification · Computed tomography · Attention-based deep neural network

1 Introduction

Biomedical imaging [1–3] has become an indispensable cornerstone of modern healthcare, revolutionizing the way we diagnose, monitor, and treat a wide range of medical conditions. From X-rays [4–6] and ultrasounds to magnetic resonance imaging (MRI) [7–9] and computed tomography (CT) [10–12] scans, the realm of biomedical imaging has witnessed remarkable advancements, offering an intricate view of the human body's inner workings.

In recent years, the integration of deep learning has driven the field of medical image analysis to a new level, where precision and efficiency have become the main priorities. The ability of deep learning models to decipher the complex visual information contained in medical images is unmatched by traditional methods. This has provided an opportunity to tackle a diverse range of medical challenges, ranging from early disease detection to treatment planning and monitoring.

Organ classification stands as a fundamental task in biomedical image analysis, with its practical applications extending far beyond academic exploration. It serves as a bedrock for a multitude of vital healthcare applications, including automated diagnosis, the formulation of patient-specific treatment strategies, and the optimization of clinical workflows. However, the seemingly straightforward task of classifying organs within medical images presents a nexus of challenges. The inherent variability in organ appearance, stemming from patient anatomy, disease state, and the type of imaging modality used, creates a complex scenario. The potential presence of pathology further complicates the task and makes accurate classification even trickier. Conventional organ identification techniques frequently depend on radiologists or other medical experts manually interpreting medical pictures. Despite their high level of expertise, these experts' evaluations may vary depending on subjective interpretation, experience level, and weariness. Additionally, humans are susceptible to cognitive biases that can subconsciously influence their interpretation. These biases can eventually lead to missed diagnoses or misidentification. Studies have shown that human accuracy in organ identification can vary, with rates typically ranging from 85 to 95% depending on the complexity of the images and the experience of the radiologist. For instance, in a study by Kim et al. [13], the pooled kappa coefficient (k) for interreader reliability of the LI-RADS Treatment Response algorithm was 0.70, indicating substantial agreement among readers but also underscoring the inherent limitations of human interpretation in medical imaging, as the perfect agreement was not achieved. This suggests that even with standardized systems, human errors in interpreting liver cancer treatment responses are still a significant concern.

Deep learning offers a compelling solution to overcome these limitations and improve the accuracy of abdominal organ identification. Massive medical image datasets that cover a variety of anatomical variances, imaging modalities, and even pathological presentations are used to train artificial neural networks. The models can acquire unique patterns that differentiate various organs through this training, which has multiple benefits like lowering human error, increasing accuracy, and enhancing generalizability.

Studies have shown that deep learning models can achieve impressive accuracy in abdominal organ identification tasks. For instance, in a recent study by Yang et al. [14], the DL model outperformed contemporary methods with an average Dice Similarity Coefficient (DSC) of 87.72 on pancreas image segmentation of CT images. In another study by Liu et al. [15], deep learning-based multi-organ segmentation methods have significantly outperformed traditional approaches, with a focus on full and imperfect annotation techniques. The study highlights advancements in network architecture, dimensions, dedicated modules, and loss functions, as well as new challenges and trends in the field.

The classification of abdominal organs from 2D CT images, as exemplified in the Organ [16] datasets, exemplifies the confluence of these challenges. These datasets offer a diverse scenario for the evaluation of such deep learning models.

In this paper, we introduce a novel attention-based deep neural network approach that exceeds the existing benchmark accuracy for the corresponding dataset. We have integrated specialized squeeze attention mechanisms into an established deep-learning architecture, Xception. The addition of the Squeeze and Excitation Block (SE block) enhances the models' ability to extract relevant features from complex 2D CT images. This research shows the potential of deep learning in medical image analysis, highlighting the importance of organ classification in healthcare. This research aims to enhance clinical decision support, facilitate better healthcare automation, and address current challenges in medical image classification. The significant contributions of this research paper are as follows:

- Introduction of an attention-based deep neural network approach that integrates the specialized squeeze-excitation attention mechanism to the Xception deep learning model for Organ{A, C, S}MNIST datasets. Figure 1 depicts our proposed workflow.
- Demonstration of the proposed model's capability to surpass benchmark accuracies on the Organ{A, C, S}MNIST datasets within the recently developed MedMNISTv2 benchmark database.
- Illustration of the potential of deep learning with specialized attention mechanisms for medical image analysis.

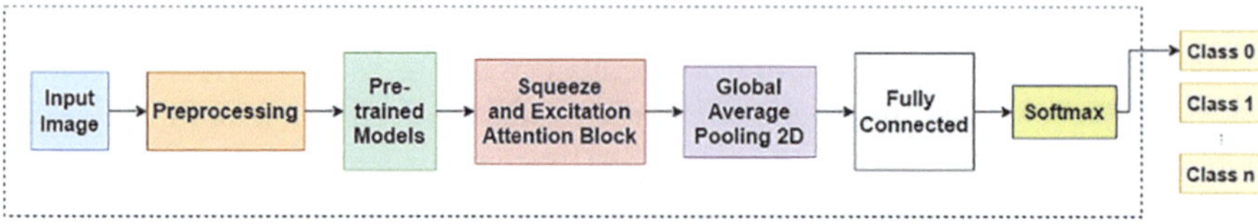

Fig. 1 Workflow for our proposed attention-based deep neural network

2 Related Study

In this section, we review some of the works performed on the previously proposed MedMNIST dataset.

In order to create appropriate CNNs for medical picture classification, LeCun et al. [17] suggested a novel context-free grammar linked to a multi-objective grammatical evolution method. The grammar allows for the inclusion of regularization layers and has layers that are highly relevant to classification problems. In order to prevent non-convergence, it also creates a search space with 6426 distinct individuals. More granularity options, ranging from 1 to 3, are offered by the gram-mar in the convolutional block creation sequence. The MedMNIST dataset is used to test the suggested method, which has never been done before.

In comparison to state-of-the-art networks and other CNNs produced by grammatical evolution, He et al. [18] demonstrate that the suggested approach produces simpler networks with comparable or better performance. The accuracy and F1-score of the networks produced by the suggested method are statistically equivalent to those of the top-ranked network across all datasets. The suggested grammar is seen as an intriguing substitute for building low-complexity, competitive CNN models for image classification tasks.

According to Li et al. [19], cost-sensitive self-paced learning (CSSPL) performs better in terms of classification accuracy and computational complexity than other automated architecture search techniques and well-designed artificial networks. Although CSSPL takes longer than well-designed neural networks, it is more useful and performs well in generalization. When compared to alternative neural architecture search algorithms and classical convolutional neural networks, CSSPL achieves the greatest results, proving its efficacy.

Self-contrastively super-vised learning (SelfCSL), a technique developed in this study by Hinton et al. [20], builds a pre-trained model via contrastive learning, which improves efficiency and stability, using data from the same domain of the current problem. The MedMNIST dataset, a collection of ten pre-processed medical open datasets, was used to test the suggested approach.

With studies examining the efficacy of representations learnt via contrastive learning on ImageNet and other vision problems, contrastive learning—as introduced by Fukushima [21]—has recently drawn more interest from a variety of academic fields.

Niu et al. [22] provides an overview of attention mechanisms in deep learning and their applications in various domains. It defines a unified model for attention structures and describes each step of the attention mechanism in detail. The authors classify existing attention models based on criteria such as the softness of attention, forms of input feature, input representation, and output representation. The authors also discuss network architectures used in conjunction with the attention mechanism and present typical applications of attention in deep learning. Furthermore, it explores the interpretability that attention brings to deep learning and presents potential future trends.

A technique called FedSLD (Federated Learning with Shared Label Distribution) was presented by Luo et al. [23] for medical research, where data privacy laws make it difficult to train machine learning models using data from different medical facilities. FedSLD addresses the issue of decentralized data across these centers in federated learning by assuming knowledge of label distributions from all participating centers. By adjusting each data sample's contribution to local optimization based on this distribution knowledge, FedSLD minimizes the impact of data heterogeneity among centers. The method is evaluated on four public image datasets with varying non-IID (non-identically distributed) data distributions, showing superior convergence performance compared to other state-of-the-art federated learning algorithms. It achieves up to a 5.50 percentage point increase in test accuracy.

Zhu et al. [24] focused on understanding the impact of various factors influencing attention mechanisms in deep neural networks. This study explores different methods of computing attention within a generalized attention framework, including Transformer attention, deformable convolution, and dynamic convolution modules. By conducting experiments across various applications, the study reveals significant insights about spatial attention in deep networks. Surprisingly, some findings challenge conventional wisdom, such as the varying importance of query and key content comparison in different attention types. For instance, it has been discovered that the comparison is less crucial for self-attention, while it's vital for encoder-decoder attention. Additionally, a strategic combination of deformable convolution with key content-only saliency proves to achieve the best balance between accuracy and efficiency in self-attention mechanisms. Overall, the results highlight the potential for enhancing attention mechanism designs in neural networks.

Zhang et al. [25] discuss the development of a novel approach called the Squeeze and Excitation Reasoning Attention Networks (SERAN) for enhancing Magnetic Resonance image super-resolution. Their SERAN approach incorporates squeeze and excitation reasoning to gather global spatial information, generating descriptors that highlight more informative regions and structures within MR images. Primitive relationship reasoning attention is introduced to establish relationships between these descriptors, refining them with learned attention. Additionally, adaptive attention vectors recalibrate feature responses, selectively utilizing global descriptors to enhance details and texture reconstruction at each spatial location. Extensive experiments demonstrate the effectiveness of SERAN, showcasing superior performance compared to existing methods quantitatively and visually on benchmark datasets. This advancement significantly improves the accuracy and quality of MR image super-resolution, offering promising potential for more reliable diagnosis and analysis in medical imaging.

Feng et al. [26] developed a new method called semi-supervised meta-learning networks (SSMN) with squeeze-and-excitation attention introduced in this paper. By using attention, the encoder can find important features and make better fault predictions. SSMN also uses unlabeled data to improve its fault recognition, even when there is not much labeled data available. A special optimizer helps make SSMN work efficiently. The method's effectiveness is proven using data from vibrations in three different bearing datasets, showing that it works well in different situations. Comparing it with other methods using the same setup, the results show that this new method is better at diagnosing faults with only a few examples to learn from.

Park et al. [27] discuss the ongoing challenges in accurately recognizing insect species, despite recent improvements using convolutional neural networks (CNNs) for fine-grained image classification. To address these challenges specific to insect recognition, the paper introduces a new network architecture. An insect dataset from the Atlas of Living Australia is used to train the model. The results show that when applied to this insect dataset, this integrated model achieves higher accuracy than numerous alternative techniques.

Lu et al. [28] developed a highly effective multilabel classification model for diagnosing various fundus diseases from color fundus images automatically. Using a convolutional neural network (CNN) enhanced with an attention mechanism, the model can accurately classify normal fundus images and seven categories of common fundus diseases. The model was trained, validated, and tested using fundus images with eight different disease labels. Performance evaluation metrics including validation accuracy, area under the receiver operating characteristic curve (AUC), and F1-score were used. Results indicate that the proposed model achieved superior performance compared to two state-of-the-art models, with a validation accuracy of 94.27%, an AUC of 85.80%, and an F1-score of 86.08%. Notably, the model showed a substantial reduction in the number of training parameters, making it computationally more efficient compared to existing models. This model presents an automated and accurate method for diagnosing multiple fundus diseases with high precision and a significantly lower computational burden. It holds promise for widespread use in large-scale screening for fundus diseases, potentially revolutionizing diagnostic processes in primary care settings.

Xu and Zhou [29] discuss the challenges posed by complex music genres and extensive music collections in retrieving music information. Manual tagging of music genres is time-consuming and resource-intensive. To address this, a new model is proposed: utilizing a convolutional neural network with a Squeeze and Excitation Block (SE-Block) for music genre classification. Bayesian optimization is employed to find the best parameters for the SE-Block. The model was tested on the GTZAN dataset, achieving an impressive classification accuracy of 92%, surpassing the performance of many previous research efforts. This approach aims to uncover hidden information within input spectrum graphs, improving the accuracy of music genre classification, and potentially enhancing music information retrieval systems.

Wu et al. [30] show the importance of image classification in addressing complex tasks like planetary exploration and unmanned driving. A subset of image classification that has drawn interest is scene image classification. Despite being well-known for its exceptional picture classification capabilities, the Xception model has not been widely applied to scene image classification. The study suggests a method utilizing transfer learning with Xception to close this gap and evaluates its effectiveness against the Inception-V3 model. The research shows that Xception-based transfer learning works better than other approaches, particularly Inception-V3, through trials on the Intel Image Classification Challenge dataset. The results show that Xception exhibits better performance, robustness, and generalization abilities, with fewer issues related to overfitting. This suggests the potential effectiveness of employing Xception for scene image classification tasks, highlighting its advantages over other models like Inception-V3.

2.1 Motivation

Our venture into the realm of medical image diagnostics is sparked by the inherent challenges present in existing diagnostic processes. Inconsistencies across manual predictions and the susceptibility to human error underscore the need for a more robust and standardized approach. Witnessing the transformative potential of deep learning in various domains, particularly in image analysis, fuelled our interest in applying this technology to healthcare. The promise of leveraging neural networks to automatically and precisely classify organs in medical images resonated strongly intending to address these critical challenges. The capacity of deep learning models to discern intricate patterns from extensive datasets offered an appealing solution to the intricacies inherent in medical images. The impetus to contribute to the development of superior diagnostic models, capable of delivering consistent and precise predictions, emerged as a guiding force. Additionally, the understanding that advancements in this area can significantly influence clinical processes, streamline diagnostics, and ultimately enhance patient care has added another dimension of encouragement. This effort is rooted in the belief that harnessing the power of deep learning is crucial for developing a more efficient and dependable diagnostic procedure, bridging existing gaps, and elevating the benchmarks of healthcare moving forward.

3 Datasets Used in Our Research Work

Our research leverages the MedMNISTv2 dataset, a versatile benchmark for 2D [31] and 3D [32] biomedical image classification. Within this comprehensive resource, our focus narrows down to three key datasets: OrganAMNIST, OrganCMNIST, and OrganSMNIST.

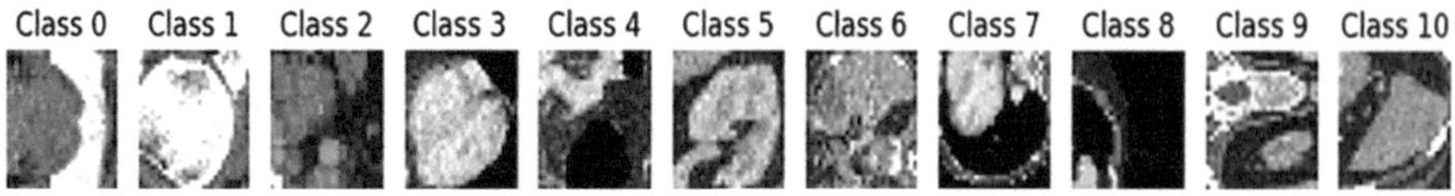

Fig. 2 Axial view images of distinct body organs in OrganAMNIST

- **OrganAMNIST**: OrganAMNIST, a pivotal element in our research, is constructed from a diverse collection of 3D CT images obtained from the Liver Tumor Segmentation Benchmark (LiTS) [33]. OrganAMNIST enables the multi-class classification of 11 distinct body organs. The data preparation begins with the conversion of Hounsfield Unit (HU) values from the 3D CT scans into grayscale images. Subsequently, 2D slices are extracted from the central regions of the 3D bounding boxes, focusing on the axial view. These 2D images are then uniformly re-sized to $1 \times 28 \times 28$ pixels dimension. Figure 2 shows the axial view images of the various organs in the dataset. The dataset includes a significant collection of 58,850 samples, divided into training, validation, and test sets, containing 34,581, 6,491, and 17,778 samples, respectively.
- **OrganCMNIST:** In our research, we also utilized the OrganCMNIST dataset, which complements OrganAMNIST and is derived from the same LiTS source. This dataset emphasizes the coronal perspective of 2D images from abdominal CT scans. OrganCMNIST is also used for multi-class classification tasks, involving the identification of the same 11 body organs. Figure 3 illustrates the coronal view images of the organs in the dataset. It comprises 23,660 samples, which are processed similarly to OrganAMNIST, including the conversion of HU [34] values into grayscale and resizing to $1 \times 28 \times 28$ pixels. The dataset is thoughtfully divided into training, validation, and test sets, with 13,000, 2,392, and 8,268 samples, respectively.
- **OrganSMNIST**: The OrganSMNIST dataset aligns with OrganAMNIST and OrganCMNIST, originating from LiTS. It is designed for the sagittal view of 2D images from abdominal CT scans. These grayscale images, each measuring $1 \times 28 \times 28$ pixels, are a valuable resource for multi-class classification tasks concerning 11 different body organs. The pre-processing includes converting HU values into grayscale and resizing uniformly. Figure 4 shows the sagittal view images of the organs in the dataset. The dataset contains a total of 25,221 samples, partitioned into training, validation, and test sets, with 13,940, 2,452, and 8,829 samples, respectively.

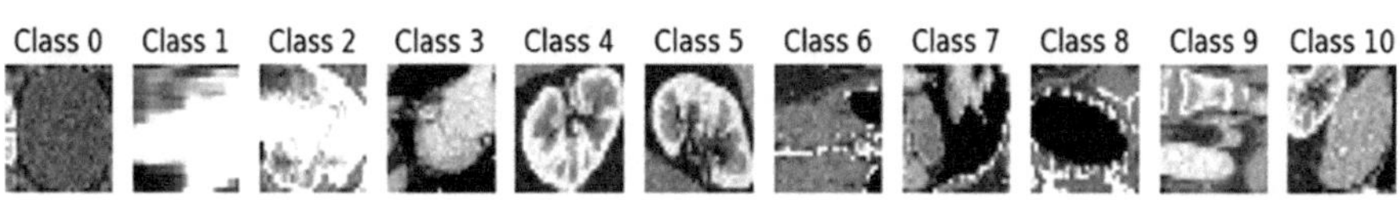

Fig. 3 Axial view images of distinct body organs in OrganCMNIST

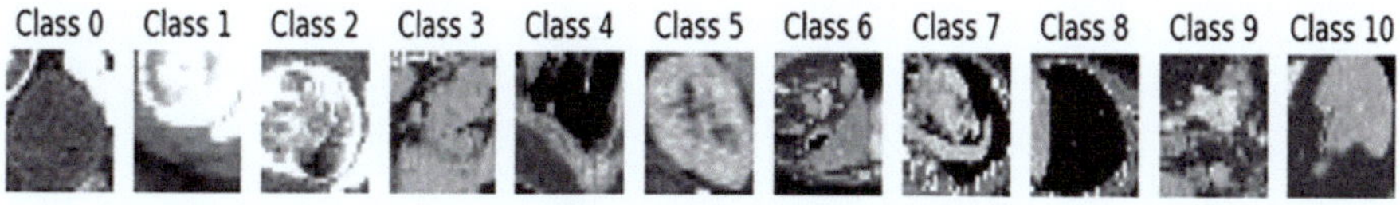

Fig. 4 Axial view images of distinct body organs in OrganSMNIST

4 Methodology and Implementation

4.1 Data Split

The dataset that has been worked upon in this study was pre-divided into training, validation, and test sets by the dataset owners. This study used the same divided train/test/validation sets to maintain standardization and enable accurate comparison with existing benchmarks. For the OrganAMNIST dataset, out of a total of 58,830 images, 34,561 are used for training, 6,491 for validation, and 17,778 for testing. Similarly, the OrganCMNIST dataset, containing 23, images, has been split into 12,975 for training, 2,392 for validation, and 8,216 for testing. The OrganSMNIST dataset, with a total of 25,211 images, is divided into 13,932 for training, 2,452 for validation, and 8,827 for testing. Figures 5, 6 and 7 shows the class distribution in train, validation and test sets for OrganAMNIST, OrganCMNIST, and OrganSMNIST datasets respectively.

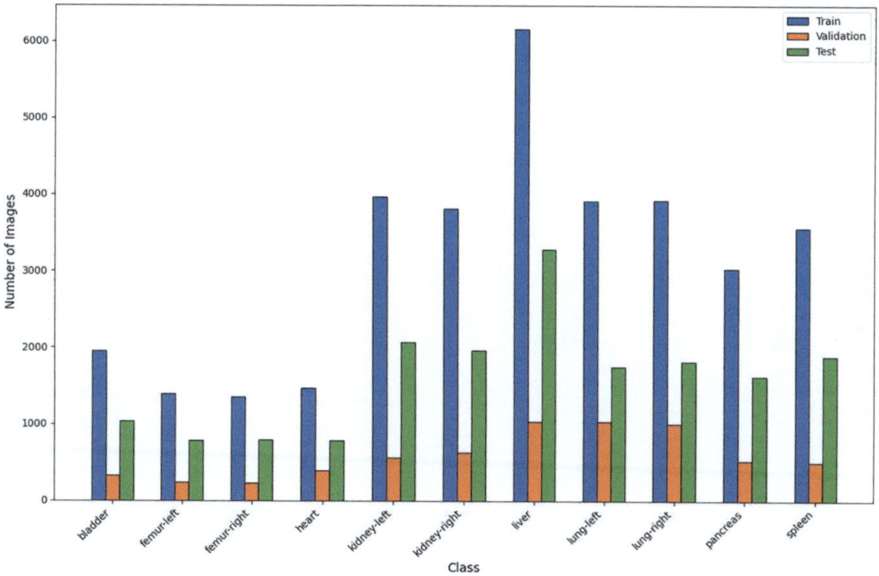

Fig. 5 Class distribution in train, validation, and test sets for OrganAMNIST

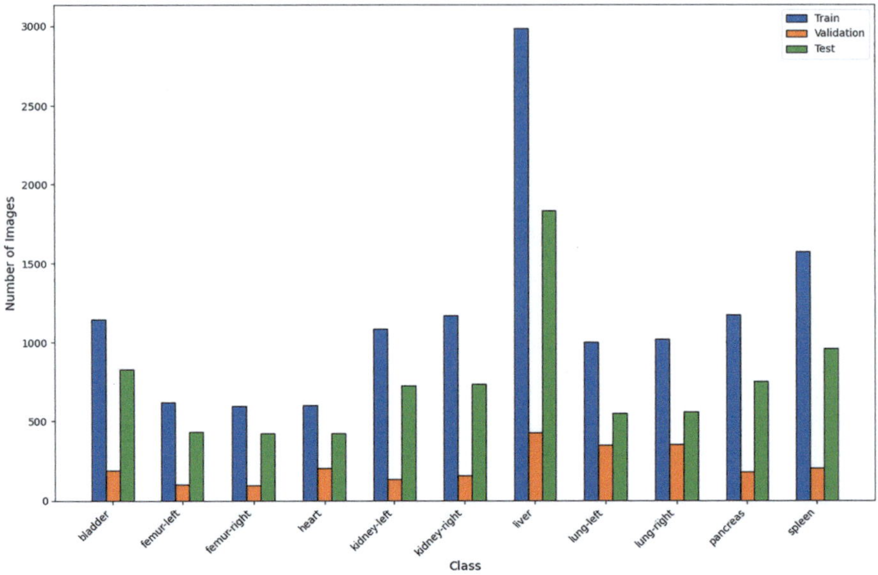

Fig. 6 Class distribution in train, validation, and test sets for OrganCMNIST

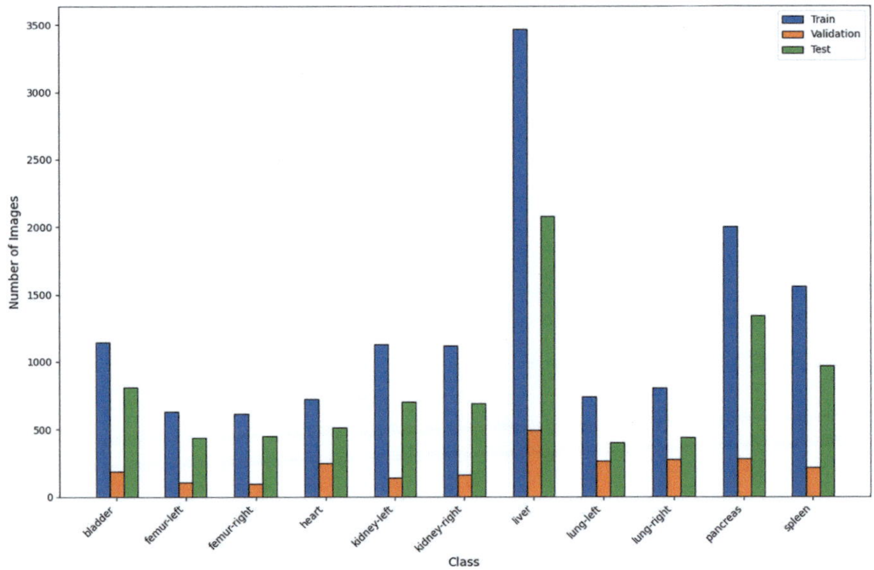

Fig. 7 Class distribution in train, validation, and test sets for OrganSMNIST

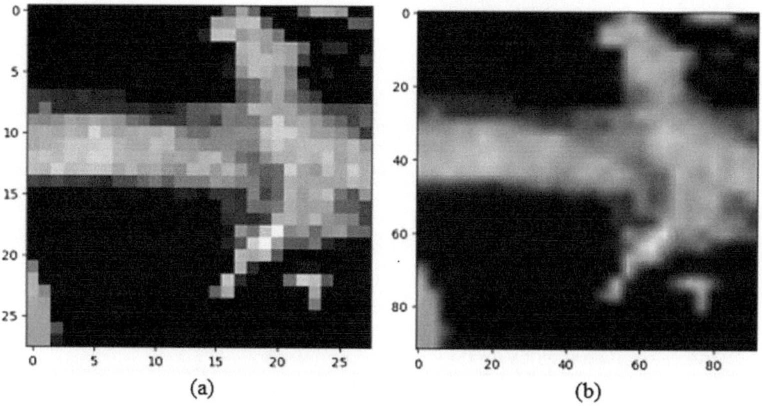

Fig. 8 **a** Original 28 × 28 image, **b** pre-processed 92 × 92 image

4.2 Data Pre-processing

Data pre-processing [35] is the initial step in training deep learning models. It plays a crucial role in configuring the data into a suitable format for comprehensive analysis. This process is essential as it involves custom modifications to raw data to uncover valuable insights and identify underlying patterns. It is important to note that all methods were carried out in accordance with relevant guidelines and regulations.

In the pre-processing of the Organ{A, C, S}MNIST datasets, the initial 28 × 28 grayscale images, as shown in Fig. 8a underwent a transformation to a detailed 92 × 92 pixel format, strategically enhancing resolution for improved model performance. Figure 8b displays the pre-processed 92 × 92 × 3 image. The process included the integration of a Squeeze-and-Excitation (SE) block, pivotal for capturing nuanced patterns. Notably, the transition from grayscale to 3-channel 92 × 92 pixel images enriched the dataset with spatial and color information, strategically expanding the feature space. This augmentation is vital for medical imaging, enhancing the dataset's potential for robust model training, especially in classifying 11 body organs from 3D CT scans, where color nuances convey diagnostic significance.

4.3 Proposed Network Architecture

The core of the proposed network is rooted in the well-established Xception [36] model that was selected for the Organ{A, C, S}MNIST dataset. The Xception model is trained on the datasets from scratch and has been configured to handle 3-channel images with an input shape of (92, 92, 3). This configuration facilitates the adept capture of intricate features crucial for organ classification across 11 distinct categories. A noteworthy augmentation to this architecture is the integration of the

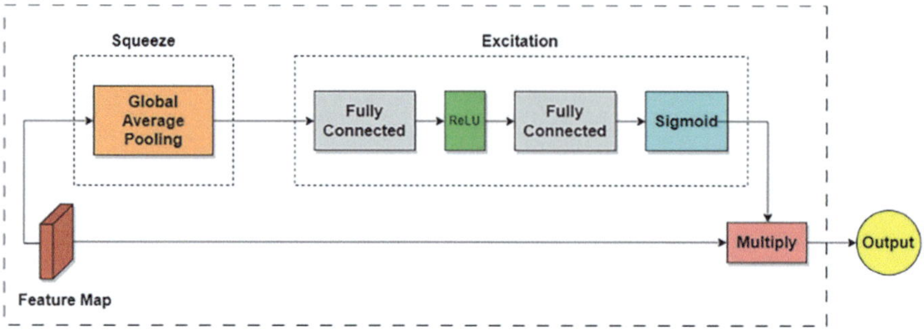

Fig. 9 Squeeze—Excitation block used in the proposed work

Squeeze-and-Excitation [37] (SE) block, strategically designed to enhance feature extraction. The SE block is a vital addition, introducing a global average pooling layer followed by reshaping and multiple dense layers featuring ReLU and Sigmoid activations. This meticulous process recalibrates features precisely, introducing a nuanced level of adaptability in feature representation. The classifier head refines these recalibrated features with a global average pooling layer, a dense layer with 1024 units and Sigmoid activation, and a final dense layer with 11 units utilizing Softmax activation for multi-class classification. Figure 9 visually depicts the intricate mechanism of the squeeze-excitation block, showcasing its global average pooling layer, reshaping, and the sequence of dense layers. Notably, this strategic combination of the Xception model and the SE block emphasizes the potential significance of the SE block in achieving precise and efficient organ classification in the domain of medical imaging. Furthermore, it's crucial to highlight that all layers of the Xception model, including the SE block, undergo fine-tuning to cater specifically to the intricacies of organ classification. Figure 10 provides a visual representation of the overall architecture, depicting both the Xception base model and the integrated Squeeze-Excitation block, showcasing their collaborative role in the network for optimal performance in organ image classification.

5 Experimental Setup and Evaluation

5.1 Experimental Setup

In the pursuit of optimal model performance and enhanced diagnostic accuracy, the experimental setup is meticulously designed to address the unique demands of the Organ {A, C, S}MNIST datasets. The key parameters and strategies include:

Epochs [38] **and Batch Size**: In the experimental design, training was capped at 100 epochs for a balance between convergence and efficiency. On top of this, an early stopping mechanism was employed to prevent overfitting. This mechanism

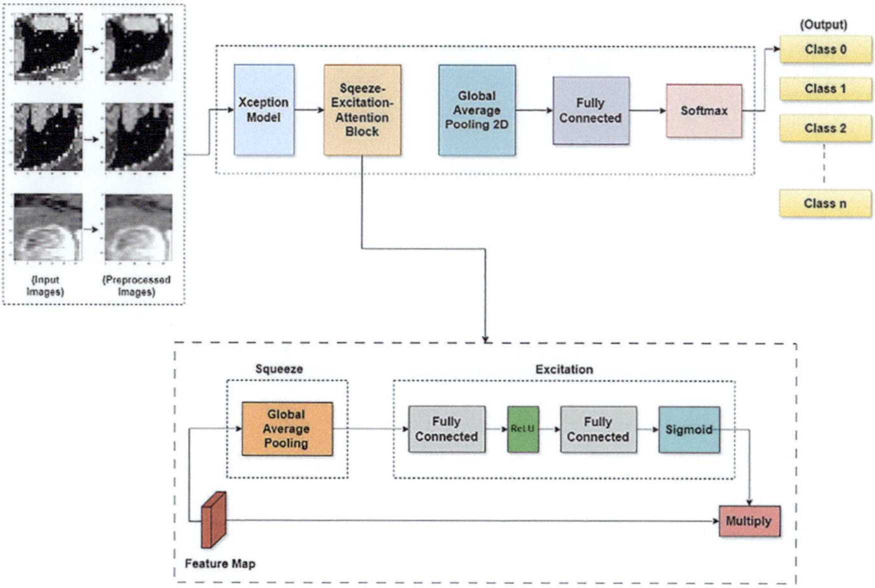

Fig. 10 Schematic diagram representing the Xception + Squeeze and Excitation block network architecture

monitors the validation performance and stops training if no improvement is observed for a predefined number of epochs, ensuring that the model does not overfit the training data. A batch size [39] of 32 optimized GPU memory usage, promoting efficient parallel processing in organ classification tasks for enhanced precision and computational efficiency.

Learning Rate Scheduler [40]: The setup utilizes a learning rate scheduler call back that dynamically adjusts the learning rate during training based on observed validation loss. The 'val_loss' parameter monitors the model's performance on the validation set, while factors like reduction extent and patience help optimize adaptability, mitigate overfitting, and contribute to improved convergence and performance.

Training from Scratch: The experimental setup entails training the model from scratch, initializing a pre-trained base model. This involves retraining all layers, including pre-trained ones, to adapt to dataset-specific features. The trainable status of the base model's layers allows the learning of dataset-specific patterns. Custom layers are then added for the organ classification task, ensuring the model captures relevant features for enhanced diagnostic accuracy on medical imaging datasets.

Early Stopping [41]: The early Stopping is a pivotal element in the experimental setup, acting as a safeguard against overfitting and bolstering model generalization. Monitoring validation loss during training interrupts the process if no improvement occurs for a set of 5 epochs. By restoring weights from the epoch with the lowest validation loss, it avoids unnecessary training iterations, conserving computational resources and ensuring the model's ability to generalize well on unseen data. The implementation has been carried out on the Tensorflow Keras framework, chosen for its simplicity and flexibility to work with.

5.2 Evaluation

Accuracy: Accuracy stands as a fundamental metric crucial for assessing the holistic performance of a model. It serves as a key indicator of the model's proficiency in making precise predictions, particularly vital in classification domains. This metric is computed as the ratio of correctly predicted instances (true positives and true negatives) to the total instances and provides a comprehensive overview of the model's overall correctness.

$$\boldsymbol{Accuracy\ Score} = (\boldsymbol{TP} + \boldsymbol{TN})/(\boldsymbol{TP} + \boldsymbol{TN} + \boldsymbol{FP} + \boldsymbol{FN}) \tag{1}$$

Precision and Recall [42]: Precision and recall metrics assume a pivotal role in evaluating the diagnostic prowess of models. Precision measures the accuracy with which models classify specific conditions, emphasizing the importance of correctly identified instances. In contrast, recall assesses models' ability to comprehensively capture instances of specific conditions. These metrics collectively provide nuanced insights into the models' diagnostic precision. It's imperative to consider precision and recall alongside accuracy, offering a comprehensive understanding of a model's performance, especially when addressing potential class imbalances or the varying costs of misclassifications.

$$\boldsymbol{Precision} = \boldsymbol{TP}/(\boldsymbol{TP} + \boldsymbol{FP}) \tag{2}$$

$$\boldsymbol{Recall} = \boldsymbol{TP}/(\boldsymbol{TP} + \boldsymbol{FN}) \tag{3}$$

F1 Score: The F1 score is a pivotal metric that provides a nuanced assessment of model performance by harmonizing precision and recall. This balance ensures a more robust understanding of a model's diagnostic accuracy. F1 scores are particularly informative when dealing with imbalanced datasets or scenarios where misclassifications carry varying consequences. These are incorporated into evaluations for a more comprehensive and nuanced analysis of model effectiveness.

$$\boldsymbol{F}1 = 2\boldsymbol{PR}/(\boldsymbol{P} + \boldsymbol{R}) \tag{4}$$

Table 1 Overall performance report produced by the proposed attention-based deep neural network on OrganAMNIST, OrganCMNIST, and OrganSMNIST datasets taken from MedMNISTv2

Dataset name	Model name	Image size, channels	Accuracy (%)
OrganAMNIST	SE block + Xception	92 × 92 × 3	96.20
OrganCMNIST	SE block + Xception	92 × 92 × 3	93.43
OrganSMNIST	SE block + Xception	92 × 92 × 3	82.95

6 Results Analysis

In this section, we present a comprehensive examination of the results obtained through the proposed deep learning model for Organ{A, C, S}MNIST datasets. Our investigation encompasses an intricate analysis of model performance and accuracy. The model exhibited a commendable balance between accuracy and computational efficiency, forming the cornerstone of our research in organ classification. Table 1 shows the overall accuracy report produced by the proposed attention-based deep neural network on Organ{A, C, S}MNIST datasets.

OrganAMNIST Evaluation: In the evaluation of OrganAMNIST, the proposed model demonstrated remarkable improvements in accuracy. The proposed model was evaluated on 92 × 92 color images which resulted in an impressive accuracy of 96.20%. Figure 11 shows the Precision, Recall and f1-scores obtained during the model's evaluation on the dataset while Fig. 12 displays the confusion matrix obtained. The accuracy curve and the loss curves of the trained model are given in Figs. 13 and 14 respectively. The model trained for 8 epochs out of the set limit of 100 and is stopped by the Early stopping. From the metrics, it can further be seen that the proposed attention-based deep neural network model demonstrated strong comparable accuracies across multiple organ classes. However, the kidney-left class came out with the lowest scores. Table 2 shows the results of 5-fold cross-validation produced by the proposed attention-based deep neural network on the OrganAMNIST dataset. The mean test accuracy came to be around 92.7% with a standard deviation of about 2.72%. This suggests that the model's performance is stable and consistent across different folds of the data.

OrganCMNIST Evaluation: The evaluation of OrganCMNIST demonstrated notable advancements in model accuracy. Our proposed approach, enhanced by SE attention blocks achieved an impressive accuracy of 93.43%, when evaluated on the dataset. Figure 15 depicts the values of Precision, Recall, and f1-scores obtained whereas Fig. 16 displays the confusion matrix obtained by the proposed attention-based deep neural network. The accuracy curve and the loss curves of the model are given in Figs. 17 and 18 respectively. The model trained for 8 epochs out of the set limit of 100 and is stopping by the Early Stopping mechanism. The results shows notable progress in accuracy improvement with lower-resolution images, crucial in resource-constrained environments. Classes such as femur-left, kidney-left, and kidney-right exhibited relatively lower scores, while liver and lungs—left and right classes, respectively, achieved the highest metrics. The reduced performance might

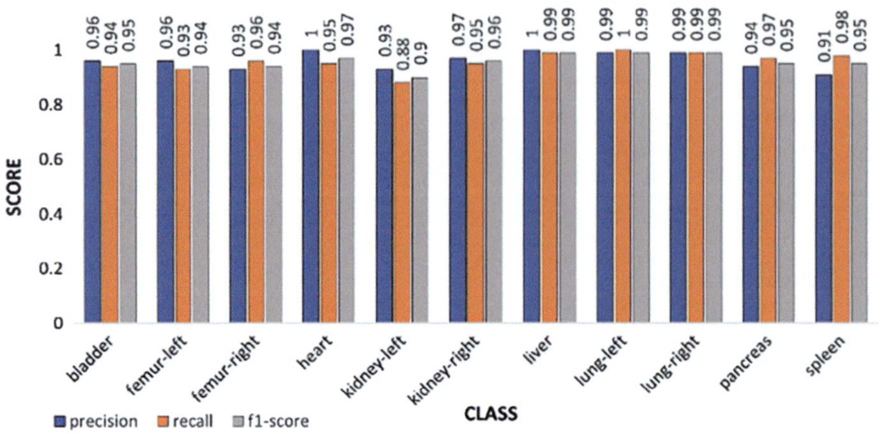

Fig. 11 Bar chart displaying the results based on precision, recall and f1-score for different classes in the OrganAMNIST dataset

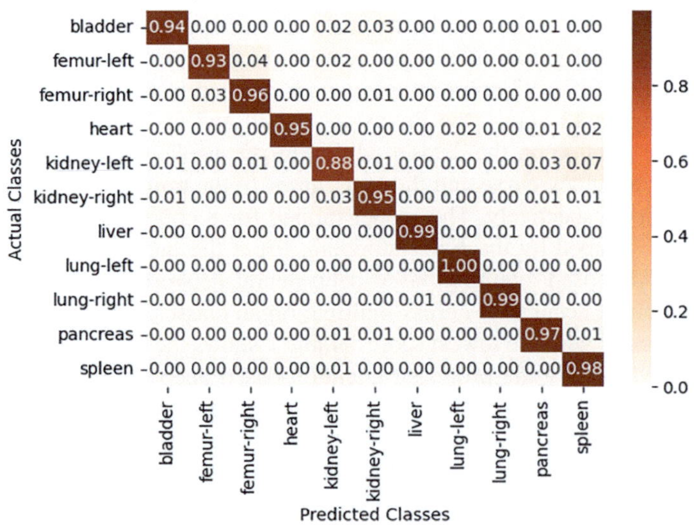

Fig. 12 Confusion matrix obtained by the proposed attention-based deep neural network for OrganAMNIST dataset

be attributed to the inherent similarity in the imaging characteristics between left and right kidney classes. The model also underwent 5-fold cross-validation, resulting in the following test accuracies for each fold as seen in Table 3. The mean cross-validation accuracy came out to be approximately 0.9702, indicating that the model performs consistently well across different subsets of the data. The standard deviation of 0.0055 signifies low variability in the performance, which suggests that the model is robust and reliable.

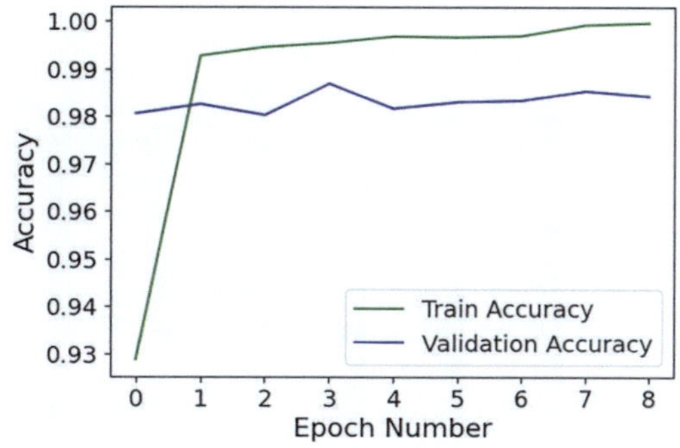

Fig. 13 Accuracy curve for trained model on OrganAMNIST

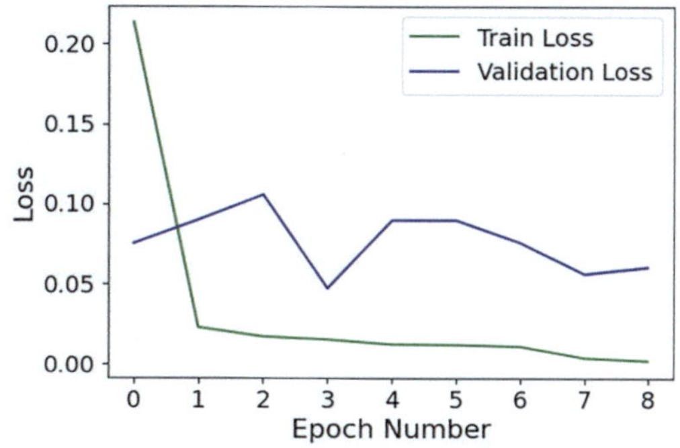

Fig. 14 Loss curve for trained model on OrganAMNIST

Table 2 Classification accuracies produced by the proposed attention-based deep neural network for each fold using 5-fold cross validation on OrganAMNIST

Fold	Test accuracy (%)
Fold 1	91.39
Fold 2	89.76
Fold 3	95.67
Fold 4	96.39
Fold 5	90.68

OrganSMNIST Evaluation: The evaluation of OrganSMNIST exhibited notable enhancement in model accuracy. The proposed approach was evaluated on 92 × 92 images and resulted in an accuracy of 82.95%. Figure 19 shows the Precision, Recall, and f1-scores obtained during the model's evaluation on the dataset. Figure 20 displays the confusion matrix obtained upon evaluation. The accuracy curve and the loss curves of the model are given in Figs. 21 and 22 respectively. The model trained for 13 epochs out of the set limit of 100 and is stopped by the Early Stopping mechanism. It is also evident that the weakest performance has been observed for femur-left and femur-right, along with kidney-left and kidney-right. The model's confusion between these pairs is apparent, highlighting the challenge posed by the extensive similarities in their images. The 5-fold cross-validation was also employed,

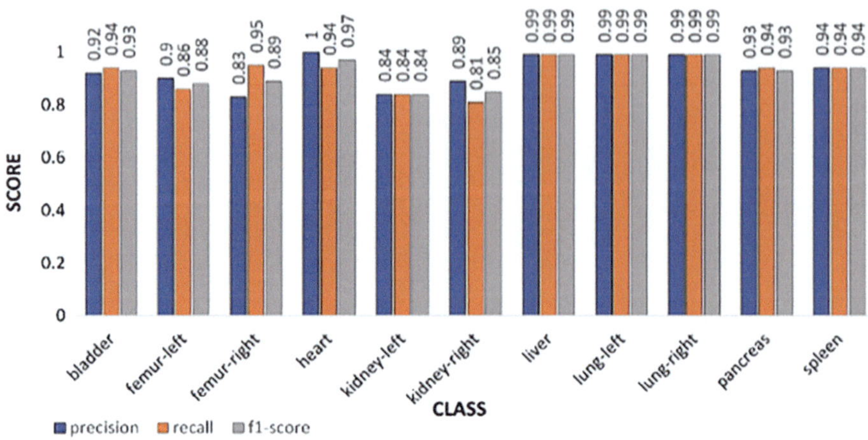

Fig. 15 Bar chart displaying the results based on precision, recall and f1-score for different classes in the OrganCMNIST dataset

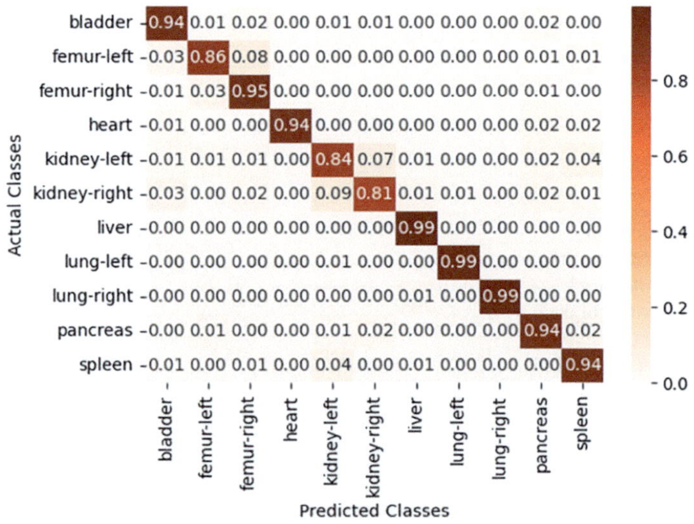

Fig. 16 Confusion matrix obtained by the proposed attention-based deep neural network for OrganCMNIST dataset

the result of which is depicted in Table 4.The mean test accuracy across the 5 folds is 75.88% ± 4.14%. This indicates that, on average, the model performs with an accuracy of approximately 75.88%, with a variability of 4.14% across different folds.

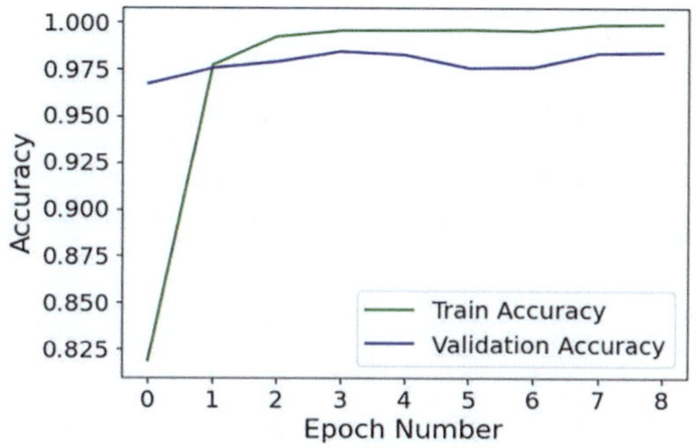

Fig. 17 Accuracy curve for trained model on OrganCMNIST

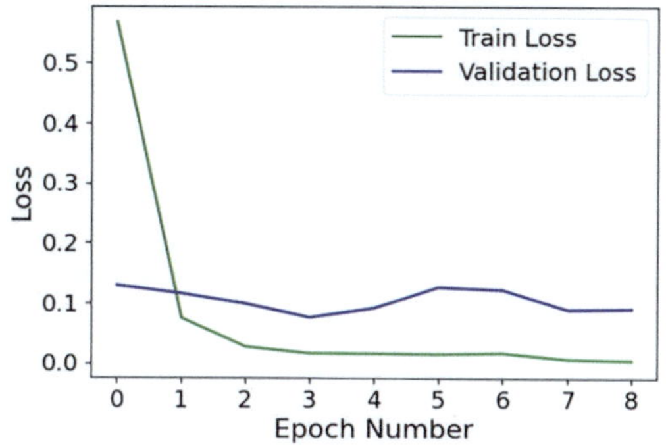

Fig. 18 Loss curve for trained model on OrganCMNIST

Table 3 Classification accuracies produced by the proposed attention-based deep neural network for each fold using 5-fold cross validation on OrganCMNIST

Fold	Test accuracy (%)
Fold 1	93.43
Fold 2	94.01
Fold 3	92.13
Fold 4	95.20
Fold 5	93.24

6.1 Comparison with Benchmark Results

The proposed approach which uses Xception model architecture further enhanced with the Squeeze and Excitation Network attention mechanism significantly outperformed the benchmark accuracies on Organ{A, C, S}MNIST datasets, showcasing notable progress in accuracy improvement with lower-resolution images and crucial in resource-constrained environments. Table 5 compares the performance of the proposed attention-based deep neural network approach with the existing benchmark accuracies. Notably, in the OrganAMNIST, the model surpassed the existing benchmark classification accuracy of 95.1% achieved by the ResNet [43] model. In OrganCMNIST, the existing ResNet-18 model achieved an accuracy of 92% when

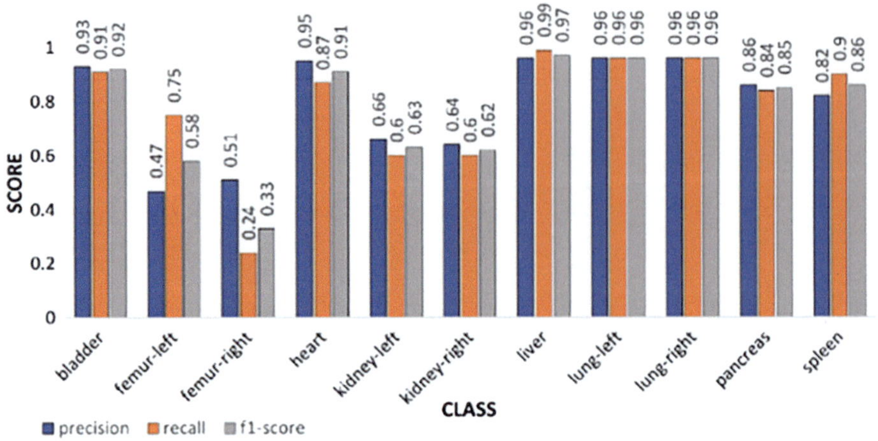

Fig. 19 Bar chart displaying the results based on precision, recall and f1-score for different classes in the OrganSMNIST dataset

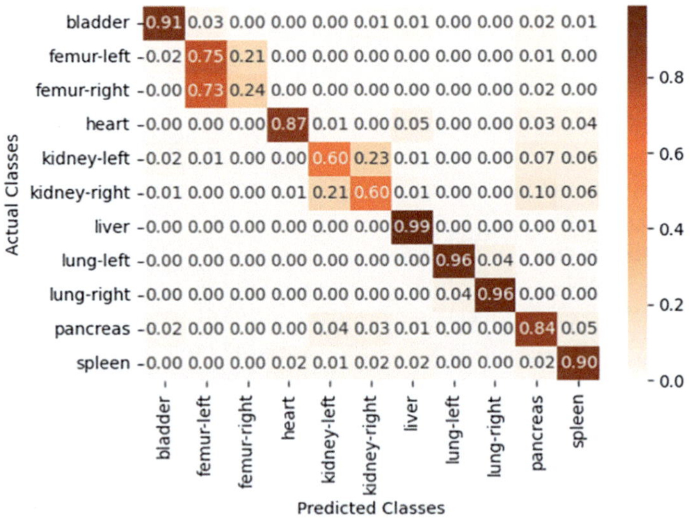

Fig. 20 Confusion matrix obtained by the proposed attention-based deep neural network for OrganSMNIST dataset

evaluated on 224 × 224 images, which has been exceeded by the proposed model's performance of 93.43%. For OrganSMNIST, the proposed model achieved 82.95%, beating the AutoKeras model's accuracy of 81.3%.

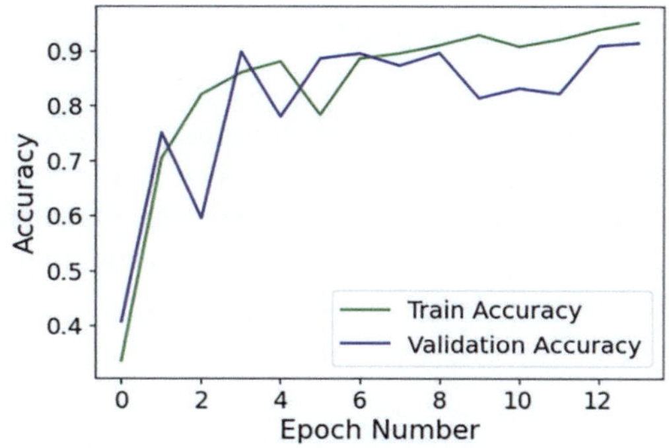

Fig. 21 Accuracy curve for trained model on OrganSMNIST

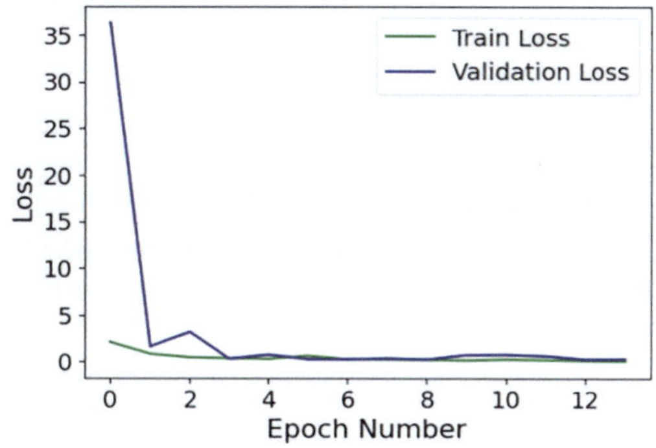

Fig. 22 Loss curve for trained model on OrganSMNIST

Table 4 Classification accuracies produced by the proposed attention-based deep neural network for each fold using 5-fold cross validation on OrganSMNIST

Fold	Test accuracy (%)
Fold 1	73.21
Fold 2	79.66
Fold 3	77.20
Fold 4	69.23
Fold 5	80.12

Table 5 Comparison of proposed attention-based deep neural network for Organ{A, C, S}MNIST dataset from MedMNISTv2

Dataset name	Authors	Approach used	Accuracy (%)
OrganAMNIST	Yang et al. [1] **Proposed method**	RestNet-18 **SE block + Xception**	95.10 **96.20**
OrganCMNIST	Yang et al. [1] **Proposed method**	RestNet-18 **SE block + Xception**	92.00 **93.43**
OrganSMNIST	Yang et al. [1] **Proposed method**	RestNet-18 **SE block + Xception**	81.30 **82.95**

6.2 Ablation Study

In this comprehensive ablation study, we delved into the nuanced effects of varying learning rates on the performance of our neural network model. The experiment focused on training a neural network architecture while manipulating the learning rates and using callbacks revealed a substantial impact on both the training dynamics and ultimate model performance.

- **OrganAMNIST**: The proposed attention-based deep neural network model underwent distinct training phases, employing fixed learning rates of 0.001, 0.0001, and 0.00001 individually. Notably, these sessions yielded accuracies of 92.16%, 94.32%, and 93.86% on the foundational Xception Model. Upon integrating an attention layer atop the base model and retraining with the same learning rates, a notable improvement ensued. The augmented architecture showcased enhanced performance, delivering accuracies of 92.56%, 95.02%, and 93.78%, respectively. Recognizing the potential for further refinement, a dynamic approach was introduced through a learning rate scheduler employed as a callback during training. This strategic addition proved effective, culminating in a noteworthy accuracy of 96.20% on the dataset when the initial learning rate was set at 0.0001.
- **OrganCMNIST**: A similar training phase was employed for this dataset. The base Xception model achieved classification accuracies of 91.48%, 92.66% and 91.40% when the learning rate was fixed at 0.001, 0.0001, and 0.00001 respectively. The Attention layer was then integrated to the existing architectures which recorded classification accuracies of 91.93%, 92.89% and 92.05% on the same set of learning rates. The addition of a learning rate scheduler proved further effective, achieving an accuracy of 93.43% on the dataset with an initial learning rate set at 0.0001. The combined impact of meticulous learning rate tuning and the adaptive scheduler significantly contributed to the heightened proficiency of the model.
- **OrganSMNIST**: The model achieved classification accuracies of 78.89%, 77.34% and 77.84% on the foundational Xception model with the learning rates fixed at 0.001, 0.0001 and 0.00001 respectively. The addition of an attention layer improved the performance of the model and it achieved classification accuracies of 81.93%, 80.56% and 80.90% on the same set of learning rates. When the model was then trained with a learning rate scheduler callback in action, the model showed even improved results, yielding an accuracy of 82.95% with the initial rate set at 0.001.

6.3 Statistical Study of Our Proposed Methodology

In this section, we used different statistical parameters to evaluate the proposed model's performance. Over the course of this study, we have employed several statistical measures to evaluate the performance of the corresponding models. Table 6 displays the result of the statistical study done on the three 2D biomedical image datasets.

- **Jaccard Index:** The Jaccard Index measures the similarity between two sets A and B by calculating the ratio of the intersection of the sets to their union. It is widely used in image segmentation tasks where it quantifies the similarity between the predicted and ground truth regions. Mathematically,

$$J(A, B) = \frac{|A \cap B|}{|A \cup B|} \tag{5}$$

  Here, |A| and |B| represent the cardinality (size) of sets A and B respectively, and ∩ and ∪ denote the intersection and union operations, respectively. The interpretation of the Jaccard Index necessitates an acknowledgment of its score range spanning from 0 to 1. A score of 0 denotes an absence of overlap, indicating suboptimal segmentation performance, as the model fails to capture any true positive pixels. Conversely, a score of 1 signifies perfect overlap, illustrating the optimal scenario where the predicted and ground truth regions are indistinguishable, exemplifying superior segmentation performance. A significant merit of the Jaccard Index lies in its inherent scale-invariant property, facilitating adaptability to diverse image sizes and resolutions. Its simplicity contributes to facile interpretation, with higher scores consistently correlating with enhanced segmentation accuracy. Furthermore, the applicability of the Jaccard Index extends to multi-class segmentation, allowing for independent assessment of each class with subsequent averaging of results. However, the Jaccard Index is not without limitations. It may exhibit sensitivity to imbalanced datasets, particularly when one class dominates the other. To make the scores unbiased, we have employed the mean-weighted Jaccard Index that assigns weight to classification labels based on the data samples in the training set.
- **Dice Similarity Coefficient**: The Dice Similarity Coefficient (DSC) measures the similarity between two sets A and B, emphasizing the balance between precision and recall. It is particularly suitable for imbalanced datasets. Mathematically,

$$Dice(A, B) = \frac{2 \times |A \cap B|}{|A| + |B|} \tag{6}$$

Table 6 Statistical study of our proposed attention-based deep neural network for Organ{A,C,S}MNIST dataset from MedMNIST v2

Dataset name	Jaccardi index	Weighted mean precision score	Weighted mean recall score	Weighted mean DSC score
OrganAMNIST	0.9281	0.9624	0.9620	0.9619
OrganCMNIST	0.8867	0.9378	0.9378	0.9375
OrganSMNIST	0.7348	0.83001	0.8298	0.8255

where, A and B are the sets being compared, $|\cdot|$ represents the cardinality of a set, and $\cap$ represents the intersection of sets. It is a critical metric for assessing image segmentation performance, with a scale from 0 to 1. A DSC of 0 signifies no overlap between predicted and ground truth regions, indicating suboptimal segmentation, while a DSC of 1 represents perfect alignment. High DSC scores indicate accurate segmentation, reflecting a strong overlap with ground truth. Conversely, lower scores suggest challenges in segmentation accuracy, potentially due to misidentification, false positives/negatives, or difficulties with nuanced boundaries. DSC's sensitivity to precision and recall balance makes it valuable in addressing imbalances within datasets.

To ensure that the proposed model utilizes adequate image features instead of noise, we conducted extensive statistical tests. The high values of the Jaccard Index, Precision, Recall, and Dice Similarity Coefficient (DSC) across all datasets, as evident above indicate that the model is accurately capturing and utilizing meaningful image features, rather than being influenced by noise.

7 Conclusions and Future Research

Our exploration into enhancing the classification of organ medical images delved into the realm of deep learning models, leading us to select Xception for its exceptional feature extraction capabilities and the balanced trade-off between computational efficiency and diagnostic accuracy. Focused on three pivotal datasets—Organ{A, C, S}MNIST, each presenting unique challenges, our study systematically evaluated these models across varied image resolutions, shedding light on the advantages of higher resolutions within our computational constraints. In our assessment of OrganAMNIST, our attention-based deep neural network, enhanced with the Squeeze and Excitation (SE) Block on 92×92 color images, achieved an impressive accuracy of 96.20%, surpassing the benchmark accuracy of 95.1%. Similar advancements were observed in OrganCMNIST, where an accuracy of 93.43% was achieved, outperforming the benchmark accuracy of 92%. Examining OrganSMNIST, we noted an accuracy of 82.95%, overshadowing the benchmark of 81.3%. Our research underscores the critical impact of model selection, image resolution choices, and the incorporation of SE Block in advancing biomedical image classification. Beyond enhancing diagnostic accuracy, our findings provide resource-efficient and effective solutions applicable to diverse healthcare contexts. The insights derived from Organ{A, C, S}MNIST datasets position these advancements as significant contributions to the broader realm of medical image analysis.

The trajectory of future research in organ classification using deep learning holds significant promise and potential for substantial improvement. As we move forward, a critical requirement is the acquisition of more extensive and diverse medical datasets to rigorously test and validate the robustness of the proposed models. Moreover, there is a compelling need for the development of additional datasets with a focus on specific medical imaging modalities to ensure comprehensive model evaluation. There is also potential in integrating multi-modal data, combining CT scans with complementary imaging modalities to broaden the scope of organ classification models. Additionally, refining transfer learning strategies could optimize the adaptation of pre-trained models to the nuanced characteristics of medical imaging datasets. Addressing the attention model architecture, more refinements and innovations in the attention blocks can be explored. This involves investigating novel attention mechanisms, experimenting with attention fusion strategies, and optimizing the block's integration within diverse deep-learning architectures. Achieving a deeper understanding of the interplay between attention mechanisms and organ-specific features can pave the way for more effective and nuanced organ classification models.

Experiment: All experiments and methods were carried out in accordance with relevant guidelines and regulations.

Informed consent was obtained from all subjects and/or their legal guardian(s).

Acknowledgements We thank all colleagues and researchers involved in this research for their valuable group discussions and suggestions on this study.

Funding The study was supported by the Russian Science Foundation (Project No. 23–71-01084).

Data Availability No datasets are generated during the current study. The datasets analyzed during this work are made publicly available in this published article.

Conflicts of Interest All the authors declare no conflict of interest.

References

1. Ghalati MK, Nunes A, Ferreira H, Serranho P, Bernardes R (2021) Texture analysis and its applications in biomedical imaging: a survey. IEEE Rev Biomed Eng 15:222–246
2. Nazir S, Dickson DM, Akram MU (2023) Survey of explainable artificial intelligence techniques for biomedical imaging with deep neural networks. Comput Biol Med 106668
3. Chakraborty S, Mali K (2023) An overview of biomedical image analysis from the deep learning perspective. In: Research anthology on improving medical imaging techniques for analysis and intervention, pp 43–59
4. Banerjee A, Sarkar A, Roy S, Singh PK, Sarkar R (2022) COVID-19 chest X-ray detection through blending ensemble of CNN snapshots. Biomed Signal Process Control 78:104000
5. Dey A, Chattopadhyay S, Singh PK, Ahmadian A, Ferrara M, Senu N, Sarkar R (2021) MRFGRO: a hybrid meta-heuristic feature selection method for screening COVID-19 using deep features. Sci Rep 11:24065
6. Banerjee A, Bhattacharya R, Bhateja V, Singh PK, Sarkar R et al (2022) COFE-Net: an ensemble strategy for computer-aided detection for COVID-19. Measurement 187:110289
7. Zhang Y, Dong Z, Wu L, Wang S (2011) A hybrid method for MRI brain image classification. Expert Syst Appl 38:10049–10053
8. Jiang S, Gu Y, Kumar E (2023) Magnetic Resonance Imaging (MRI) brain tumor image classification based on five machine learning algorithms. Cloud Comput Data Sci 122–133
9. Jiang H, Diao Z, Shi T, Zhou Y, Wang F, Hu W, Zhu X, Luo S, Tong G, Yao YD (2023) A review of deep learning-based multiple-lesion recognition from medical images: classification, detection and segmentation. Comput Biol Med 106726

10. Raj R, Siironen J, Skrifvars MB, Hernesniemi J, Kivisaari R (2014) Predicting outcome in traumatic brain injury: development of a novel computerized tomography classification system (Helsinki computerized tomography score). Neurosurgery 75:632–647
11. Gupta K, Bajaj V (2023) Deep learning models-based CT-scan image classification for automated screening of COVID-19. Biomed Signal Process Control 80:104268
12. Kundu R, Basak H, Singh PK, Ahmadian A, Ferrara M, Sarkar R (2021) Fuzzy rank-based fusion of CNN models using Gompertz function for screening COVID-19 CT-scans. Sci Rep 11:14133
13. Kim DW, Choi SH, Lee JS, Kim SY, Lee SJ, Byun JH (2021) Interreader reliability of liver imaging reporting and data system treatment response: a systematic review and meta-analysis. Br J Radiol
14. Yang Z, Zhang L, Zhang M, Feng J, Wu Z, Ren F, Lv Y (2019) Pancreas segmentation in abdominal CT scans using inter-/intra-slice contextual information with a cascade neural network 2019
15. Liu X, Qu L, Xie Z, Zhao J, Shi Y, Song Z (2023) Towards more precise automatic analysis: a comprehensive survey of deep learning-based multi-organ segmentation. arXiv:2303.00232
16. Yang J, Shi R, Ni B (2021) MedMNIST classification decathlon: a lightweight autoML benchmark for medical image analysis. In Proceedings of the 2021 IEEE 18th International Symposium on Biomedical Imaging (ISBI). IEEE, pp 191–195.
17. LeCun Y, Bengio Y, Hinton G (2015) Deep learning. Nature 521:436–444
18. He K, Zhang X, Ren S, Sun J (2016) Deep residual learning for image recognition. In: Proceedings of the IEEE conference on computer vision and pattern recognition, pp 770–778
19. Li H, Li J, Zhao Y, Gong M, Zhang Y, Liu T (2021) Cost-sensitive self-paced learning with adaptive regularization for classification of image time series. IEEE J Sel Top Appl Earth Observ Remote Sens 14:11713–11727
20. Hinton GE, Osindero S, Teh YW (2006) A fast learning algorithm for deep belief nets. Neural Comput 18:1527–1554
21. Fukushima K (1980) Neocognitron: a self-organizing neural network model for a mechanism of pattern recognition unaffected by shift in position. Biol Cybern 36:193–202
22. Niu Z, Zhong G, Yu H (2021) A review on the attention mechanism of deep learning. Neurocomputing 452:48–62
23. Luo J, Wu S (2022) Fedsld: federated learning with shared label distribution for medical image classification. In: Proceedings of the 2022 IEEE 19th International Symposium on Biomedical Imaging (ISBI). IEEE, pp 1–5.
24. Zhu, X., Cheng, D, Zhang, Z., Lin, S, Dai, J. An empirical study of spatial attention mechanisms in deep networks. In Proceedings of the Proceedings of the IEEE/CVF international conference on computer vision, 2019, pp 6688–6697.
25. Zhang Y, Li K, Li K, Fu Y (2021) MR image super-resolution with squeeze and excitation reasoning attention network. In: Proceedings of the IEEE/CVF conference on computer vision and pattern recognition, pp 13425–13434.
26. Feng Y, Chen J, Zhang T, He S, Xu E, Zhou Z (2022) Semi-supervised meta-learning networks with squeeze-and-excitation attention for few-shot fault diagnosis. ISA Trans 120:383–401
27. Park YJ, Tuxworth G, Zhou J (2019) Insect classification using Squeeze-and-Excitation and attention modules-a benchmark study. In: Proceedings of the 2019 IEEE international conference on image processing (ICIP). IEEE, pp 3437–3441.
28. Lu Z, Miao J, Dong J, Zhu S, Wu P, Wang X, Feng J (2023) Automatic multilabel classification of multiple fundus diseases based on convolutional neural network with squeeze-and-excitation attention. Transl Vis Sci & Technol 12:22–22
29. Xu Y, Zhou W (2020) A deep music genres classification model based on CNN with squeeze & excitation block. In: Proceedings of the 2020 Asia-Pacific signal and information processing association annual summit and conference (APSIPA ASC). IEEE, pp 332–338
30. Wu, X, Liu, R, Yang, H, Chen, Z. An xception based convolutional neural network for scene image classification with transfer learning. In Proceedings of the 2020 2nd international conference on information technology and computer application (ITCA). IEEE, 2020, pp 262–267.

31. De Vos BD, Wolterink JM, De Jong PA, Viergever MA, Išgum I (2016) 2D image classification for 3D anatomy localization: employing deep convolutional neural networks. In: Proceedings of the medical imaging 2016: image processing, vol 9784. SPIE, pp 517–523.
32. Korolev S, Safiullin A, Belyaev M, Dodonova Y (2017) Residual and plain convolutional neural networks for 3D brain MRI classification. In: Proceedings of the 2017 IEEE 14th international symposium on biomedical imaging (ISBI 2017). IEEE, pp 835–838
33. Bilic P, Christ P, Li HB, Vorontsov E, Ben-Cohen A, Kaissis G, Szeskin A, Jacobs C, Mamani GEH, Chartrand G et al (2023) The liver tumor segmentation benchmark (lits). Med Image Anal 84:102680
34. Razi T, Niknami M, Ghazani FA (2014) Relationship between Hounsfield unit in CT scan and gray scale in CBCT. J Dent Res Dent Clin Dent Prospect 8:107
35. Bhattacharyya S (2011) A brief survey of color image preprocessing and segmentation techniques. J Pattern Recognit Res 1:120–129
36. Chollet F (2017) Xception: deep learning with depthwise separable convolutions. In: Proceedings of the IEEE conference on computer vision and pattern recognition, pp 1251–1258
37. Hu J, Shen L, Sun G (2018) Squeeze-and-excitation networks. In: Proceedings of the Proceedings of the IEEE conference on computer vision and pattern recognition, pp 7132–7141
38. Thomas D, Maraston C, Bender R, De Oliveira CM (2005) The epochs of early-type galaxy formation as a function of environment. Astrophys J 621:673
39. Schmeiser B (1982) Batch size effects in the analysis of simulation output. Oper Res 30:556–568
40. Yedida R, Saha S (2019) A novel adaptive learning rate scheduler for deep neural networks. arXiv:1902.07399
41. Prechelt, L. Early stopping-but when? In Neural Networks: Tricks of the trade, Springer, 2002; pp 55–69.
42. Flach P, Kull M (2015) Precision-recall-gain curves: PR analysis done right. In: Advances in neural information processing systems, p 28
43. Odusami M, Maskeliūnas R, Damaševičius R, Krilavičius T (2021) Analysis of features of Alzheimer's disease: detection of early stage from functional brain changes in magnetic resonance images using a finetuned ResNet18 network. Diagnostics 11:1071

Artificial Intelligence and Machine Learning Use in Agriculture Domain: A Review

Vinay Kumar and Sushil Sharma

Abstract The integration of artificial intelligence (AI) and machine learning (ML) methods in agriculture has garnered considerable interest in recent years. It has emerged as an authoritative tool that can transform the agricultural industry by improving productivity, boosting resource consumption, and improving decision-making processes. Integrating AI and ML technologies in the agricultural supply chain is revolutionizing, the domain by bringing in robust monitoring and prediction and quick decision-making abilities. A comprehensive literature scrutiny of the applications of artificial intelligence devices and machine learning an authoritative revolutionize farming practices and improve crop yield, resource management, and sustainability in agriculture. This present study explores the various applications of AI and ML in agriculture, focusing on their benefits, challenges, and prospects.

1 Introduction

Artificial intelligence is the branch of science that deals with the development of machines to mimic human intelligence. Machine learning is a subset of artificial intelligence that empowers systems to acquire data autonomously, without the necessity for explicit programming. It is essential to differentiate between these two technologies, as they are often mistaken for one another. Deep learning technique is a specialized area within the broader field of machine learning progression, which itself falls under the umbrella of artificial intelligence. Both these latest technologies are considered subcategories within the larger realm of artificial intelligence [1]. In Fig. 1, the diagram showcases the connections between Artificial Intelligence (AI), Machine Learning (ML), and Deep Learning (DL). The visual depiction illustrates the interconnections and interfaces among these three concepts within the realms of technology and data science. There are three primary categories of machine learning: supervised learning, unsupervised learning, and reinforcement learning (Fig. 2). Supervised learning consist of training a model on labelled data,

V. Kumar (✉) · S. Sharma
Faculty of Agricultural Engineering, SKUAST-J, Jammu, Jammu and Kashmir, India
e-mail: vinaykumarmangotra27@gmail.com

S. Chand Kumain et al. (eds.), *AI and ML Techniques in Image Processing and Object Detection*, https://doi.org/10.1007/978-981-96-7445-9_2

unsupervised learning includes finding patterns in unlabelled data, and reinforcement learning encompasses training a model to make sequences of decisions in an environment to maximize a reward [2]. Each type has its applications and use cases in various industries such as healthcare, food industries, finance, and technology.

Agriculture is constantly pressed upon to yield more with less resources. AI and ML techniques can enhance resource exploitation by analysing agricultural data. These technologies assist farmers to make decisions based on data analysis, optimize crop yields, and minimize resource wastage. AI and ML algorithms can investigate

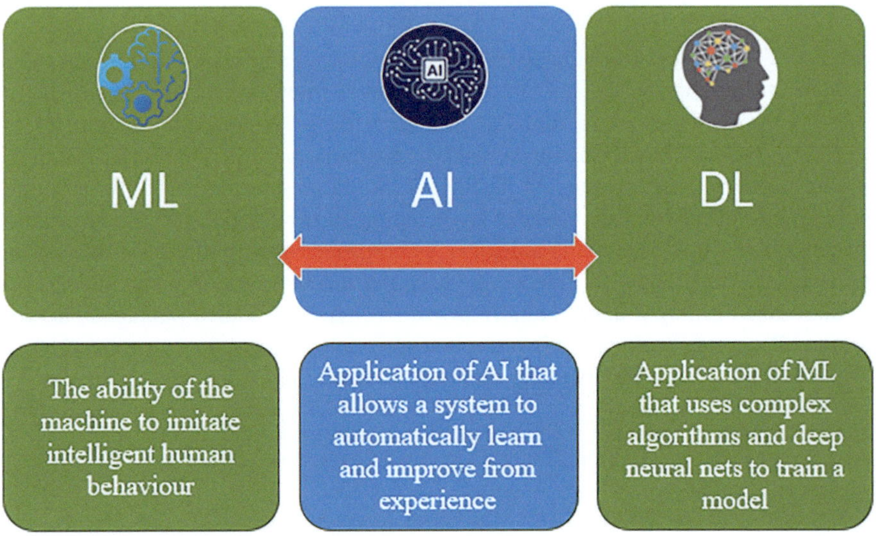

Fig. 1 The link between AI, ML, and DL

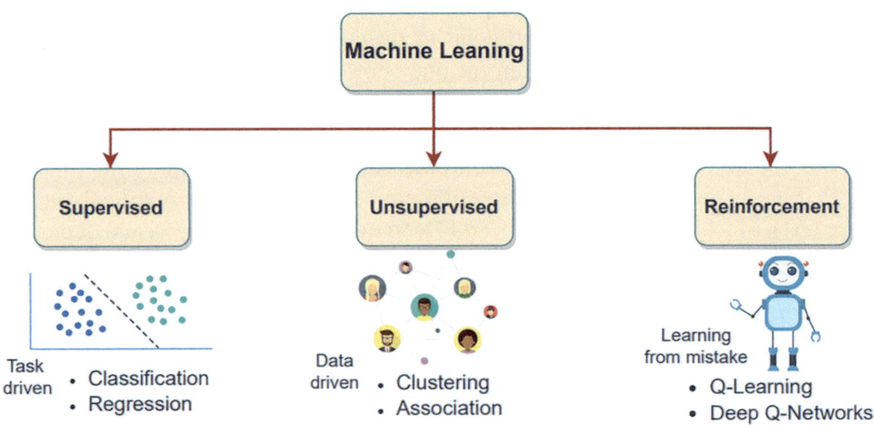

Fig. 2 Types of machine learning

enormous amounts of data to provide insights into soil health, weather patterns, and pest infestations, allowing farmers to take pre-emptive measures to ensure the health and productivity of their crops. Additionally, these technologies can also mechanise various tasks such as irrigation, harvesting, and sorting, leading to better efficiency and minimised labour costs [3]. Overall, AI and ML are transforming the agricultural landscape by empowering farmers with the tools to sort informed and sustainable choices. They are also used to develop precision agriculture techniques, such as the usage of drones and sensors to observe and accomplish crops at a granular level. This allows for targeted application of resources such as water, chemical fertilizers, and pesticides, leading to further sustainable and environmentally friendly farming practices. Additionally, these machineries can also support in predicting market demand and optimizing supply chain management, leading to better profitability for farmers [4].

In the future, AI and ML are expected to continue to play a noteworthy role in agriculture, with the potential to further improve productivity, sustainability, and resilience in the face of climate change and other challenges. In an optimal smart ecosystem, a farmer would be assisted by an artificially intelligent aide to regulate the best date and method for land preparation, based on the GIS and remote sensing data of the region. The farmer would then utilize a blockchain and recommender system-enabled supply chain to procure high-quality seeds for sowing after the land preparation. Furthermore, low-cost smart weeding and fertigation systems would manage the scheduled weeding tasks effectively. AI-powered mobile applications can effectively identify pests and diseases in crops, providing farmers with suitable management practices to combat them [5]. As these technologies continue to advance; they are expected turn out to be even more accessible and affordable for farmers of all scales, further democratizing the benefits of data-driven decision-making in agriculture [6]. Overall, the integration these technologies in agriculture embraces inordinate assurance for the future of food production and the sustainability of our planet (Fig. 3).

2 Application of Artificial Intelligence and Machine Learning in Agriculture

In the present scenario, AI and ML techniques are being exponentially applied in the various areas of the agricultural domain. The practice of AI and ML in agriculture has advanced significantly in recent years, enhancing crop monitoring in real time, yield forecast, pest detection, and soil analysis. These technologies enable data-driven decision-making, resource optimization, and improved productivity for farmers. AI and ML algorithms can scrutinise large amount of agricultural data to ascertain patterns and trends, while robotics and automation technologies reduce reliance on manual labour. Additionally, AI-powered automated robotics have been advanced to perform tasks such as planting seed, harvesting, and irrigation, reducing the reliance

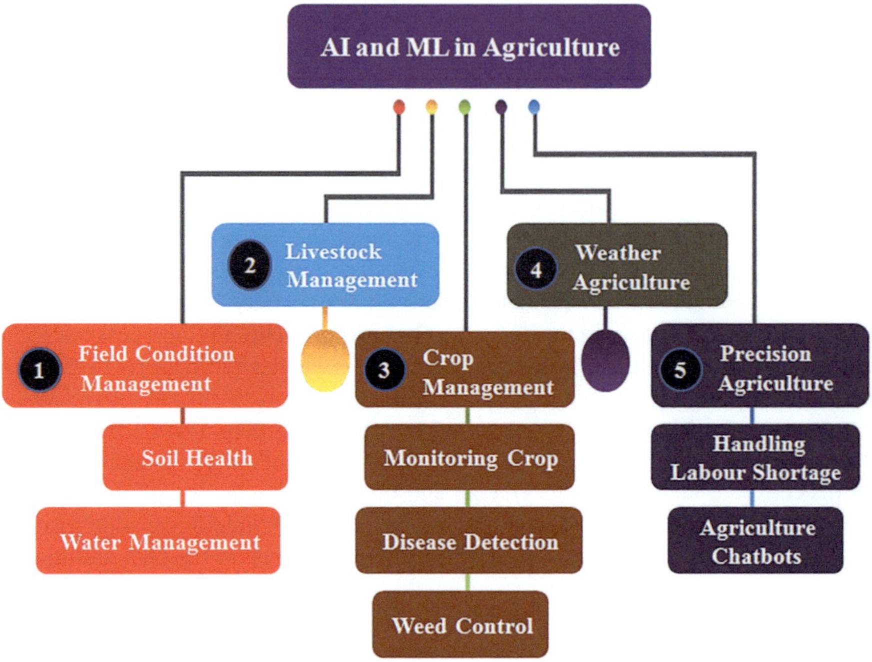

Fig. 3 AI and ML in agriculture and allied sector

on manual labour and increasing operational efficiency. The integration of AI and ML in agriculture has the capacity to transform the industry, addressing worldwide food sanctuary challenges and mitigating the impact of climate variations on agricultural production. The major applications of AI and ML based techniques on these areas are discussed in the subsequent sections.

2.1 Crop Health Management

Plant diseases can result in reduced growth and have a destructive impact on crop yields, leading to an estimated global economic loss of up to $20 billion per year. In India, the agricultural sector experiences an annual loss of 17.5 to 20%, amounting to US$ 42.66 million due to crop diseases. In many cases, the whole crop production is destroyed due to crop diseases. Conventional techniques for recognising plant' diseases are often ineffective if not applied early in the pathogenesis process when symptoms are minimal. These approaches are not capable to provide spatialized diagnostic results for plant diseases [7]. Traditional methods typically depend on experts, experience, and guides, which can be costly, time-consuming, and laborious with limited accuracy. Additionally, farmers are increasingly resorting to agrochemicals to safeguard crops against diseases, posing risks to the environment, soil, and water quality. India encounters notable issues regarding water quality, especially concerning surface and groundwater pollution (up to 70%) according to the National Water-Quality Assessment (NAWQA).

Therefore, a technologically driven agricultural revolution is crucial for solving problems and increasing efficiency at a realistic cost with minimal environmental impact. The adoption of advanced technologies like Internet of Things devices, intelligent algorithms, sensors, and modern machines has transformed agriculture. Artificial intelligence, machine learning, and deep learning machineries are being used to improve computer vision-based systems for autonomous crop disease monitoring. These technologies can autonomously extract features and provide farmers with data accuracy ranging from 85 to 95%, along with a 90% chemical use efficiency compared to traditional methods [8].

Crop management has been revolutionized by the integration of these modern sensor equipped technologies. These advanced tools have empowered farmers to take more cognisant decisions regarding planting, irrigation, pest control, and harvesting. By scrutinising massive amounts of data, AI driven algorithms can deliver valuable understandings into crop health and optimal resource allocation by identifying the trends that may not be apparent to the human eye, allowing for more efficient and sustainable agricultural practices. The most recent example of cutting-edge technology in agriculture involves the development of a Normalized Difference Vegetation Index (NDVI) sensor that utilizes AI-ML technology to monitor crop health in real-time [9]. By analysing MODIS/AwiFS satellite images, this sensor provides accurate and timely assessments of vegetation health, empowering farmers to sort cognisant decisions on crop management and optimization strategies. By training algorithms on diverse datasets, farmers can develop tailored strategies to maximize productivity and minimize waste. Many AI and ML-powered sensors can monitor crop health with real-time management strategies to protect the crop from various types of diseases (Table 1). This level of precision not only improves crop revenues but also decreases the environmental effect of farming operations.

2.2 *Soil and Irrigation Management*

Soil and irrigation are the most viable components of agriculture. The soil and irrigation are the determinant factors for the optimum crop yield. Soil's nutrients are necessary for crop growth and productivity, and meeting global food demand. Previously, soil scientists relied on labour-intensive techniques for data gathering, which were time-consuming and limited. However, AI and ML now provide powerful tools to examine enormous amounts of data and extract insights to improve soil health. These technologies can process diverse soil data to identify patterns, correlations, and anomalies that may be difficult for humans to discern, and improve our understanding of soil variation. The use of AI-driven sensors and robotics allows for the real-time supervision of soil conditions, as well as the optimization of irrigation and

Table 1 AI and ML-driven technologies for monitoring crop health

S. No.	AI-ML enabled technologies	Disease detection	Advantages
1	Digital camera (RGB)	Cotton bacterial angular, leafspots, early blight, fusarium wilt, rusts, etc	• Vegetation features can be captured in grayscale or colour images • Visible spectrum aids in better disease detection at the leaf level • Lightweight, affordable, user-friendly, simple data processing, and suitable for minimal work settings
2	Multispectral camera	Frogeye leaf spot, grapevine leaf stripe, white leaf spot, disease (GLSD), bacterial soft rot, blights, powdery mildew, viruses, anthracnose, blasts, etc	• Cost-effective pricing, fast frame capture, and enhanced durability compared to RGB cameras boost productivity • Spans electromagnetic spectrum from visible to Near-Infrared (NIR) for computing different vegetation indices • Detects and records radiations from both visible and invisible sections of the electromagnetic spectrum

(continued)

Table 1 (continued)

S. No.	AI-ML enabled technologies	Disease detection	Advantages
3	Hyperspectral sensing	Leaf hoppers, blasts, leaf rollers, nematodes, oriental fruit moths, rusts, peach twig borer, scabs, etc.	• Able to detect and document numerous narrow bands and continuous spectra • Offers researchers and farmers a more inclusiveperception into disease and crop spectral characteristics • Capable of recognizing and capturing a wide range of spectral features
4	Thermal infrared cameras, including near InfraRed, short-wave infra-red, mid-wave InfraRed, long-wave InfraRed, and far InfraRed	Recently used to monitor diseases such as smut, leaf spot sheath blight, cercospora, tungro disease, scab, stem rot of rice, mildews, grassy stunt disease, rice ragged stunt virus, etc.	• Capable of detecting infrared light, making it appropriate for use throughout both day and night • Provides a greater amount of information on plant health compared to alternative sensors • Sensitive to the infrared spectrum, enhancing its ability to gather data on plant health

fertilization methods, resulting in more efficient agricultural practices, sustainable land management, and enhanced soil health [10].

AI-integrated soil sensors require the development and implementation of algorithms, models, and systems that empower them to perceive and comprehend their surroundings, reason and make judgements based on available data, and take appropriate actions to achieve specific objectives. In contrast, ML is a sub-domain of AI that concentrates on creating algorithms or models that assist computers to study from data and forecasts results without explicit programming. ML algorithms are designed to automatically analyse large datasets, identify patterns, and extract meaningful insights to enhance their performance over time. Table 2 summarizes research on soil parameters using AI, ML, sensors, and IoT systems, providing valuable insights into advancements made by various authors in this field.

The United Nations reports that 40% of the global population resides in regions facing moderate to high water stress, with uneven distribution worldwide. Countries

Table 2 Soil and water management using AI and ML new innovative techniques

S. No.	Categories specified in the assignment	Model/Techniques/Methods used	Accuracy/Results	References
1	Soil fertility indices for pH, organic carbon, boron, phosphorus, and potassium are categorized by village	ELM uses different activation functions such as, Gaussian radial basis, hard limit triangular basis, sine-squared and hyperbolic tangent	Achieved 80% accuracy	[14]
2	Soil organic matter content, soil pH, soil temperature	Four machine learning models: cubist, ELM, LS-SVM, ANN and PLSR	The value of R2 was 0.81	[15]
3	The parameters include organic carbon (OC), total nitrogen (TN) and moisture content (MC)	Cubism, PLSR, LS-SVM, and PCA	RMSEP (MC) = 0.457%, and RPD (2.25), RMSEP (TN) = 0.071 and RPD (1.96)	[16]
4	An IoT-based solution for efficient watering in farming areas	CNN models integrated with an IoT-based system	Achieved 90% accuracy	[17]
5	Plant's water content	Reflectance across a wide spectrum of wavelengths	The most optimal models water band index (WBI), MSI, NDWI1640, and NMDI	[18]

such as India, Mexico, USA, and China are identified as top consumers of groundwater sources. Agricultural activities contribute to nearly 70% of water withdrawal, while industrial and domestic sectors account for 22 and 8% respectively. In India, agriculture alone utilizes 90% of groundwater due to excessive extraction and inefficient irrigation methods, highlighting agriculture as a major contributor to freshwater scarcity [11]. Urbanization, agricultural intensification, and climate change have led to a rise in water demand and degradation of freshwater, posing significant challenges in regions already facing water stress. Unsustainable groundwater extraction supports over 25% of the world's population and 40% of global agricultural production. By 2030, the country's water demand is projected to double the available supply. Therefore, the soil and irrigation-related issues should be managed properly and cautiously to ensure a potential yield in crops. In this regard, AI and ML-based techniques have shown the potential ability to resolve soil and irrigation-related issues in crops [12].

Advanced AI-ML technology can analyse data from satellite, plane, or drone imagery to identify irrigation issues by interpreting patterns in images using machine learning algorithms. By combining imagery with soil and plant-based sensors, real-time data can accurately determine irrigation needs and alert farmers to potential problems, enabling efficient irrigation management and sinking the risk of under or over-watering. Additionally, forecasting weather patterns helps farmers prepare for any challenges posed by nature. AI-powered technology also provides a cutting-edge approach to monitoring water quality in real-time. Utilizing state-of-the-art sensors, these systems analyse quality metrics non-stop, delivering instant updates on pH, temperature, dissolved oxygen, and pollutant levels. By integrating machine learning and data analytics, AI-enhanced systems can swiftly identify irregularities and deviations from standard conditions. They can also promptly notify decision-makers of potential water concerns, allowing organizations to take immediate action to address sources of contamination [13].

This cutting-edge technology enables farmers to make well-informed verdicts based on data, by monitoring causes such as temperature, soil moisture levels, and nutrient content. By doing so, it guarantees that resources are consumed efficiently, potentially resulting in a substantial increase in yield, possibly up to 30%. Through effective monitoring and optimal control of irrigation, this technology also facilitates significant water conservation, with possible savings reaching from 30 to 60%. It can lead to a reduction in indirect costs related to energy use, such as electricity or fossil fuel for pumping, ultimately enhancing cost-effectiveness.

2.3 *Pest Management*

Pesticides are extensively applied on a global scale, resulting in detrimental effects on both human well-being and the environment. India holds the position as the fourth-largest manufacturer of pesticides worldwide. Insecticides, fungicides, and herbicides are frequently utilized for pest management in agricultural practices. Nevertheless, insecticides constitute the largest portion of the overall pesticide usage in India. Over the last ten years, there has been a notable rise in the per-hectare consumption of pesticides in India, marking an increase of approximately 50 percent compared to the previous decade. A recent study published in Nature Geoscience revealed that approximately 385 million individuals working in the agricultural sector experience acute pesticide poisoning on an annual basis [19]. The symptoms linked with this form of poisoning vary from weakness and headaches to vomiting, skin rashes, and even failure of vital organs such as the heart, lungs, or kidneys, as well as disorders of the nervous system. Shockingly, the study also establish that around 11,000 people succumb to acute pesticide poisoning each year, excluding cases of suicide related to pesticide exposure.

A recent study has also exposed that a significant 64 percent of agricultural soil worldwide is contaminated with pesticide residues, leading to the widespread issue of global pesticide pollution. Conventional techniques for uniformly applying pesticides across fields lead to excessive chemical usage, significant wastage (70–90%), higher cultivation expenses, soil contamination, water pollution, environmental degradation, and adverse effects on farm health. Consequently, it is imperative to implement novel techniques for pesticide application to mitigate the detrimental effects it poses to both human well-being and the environment [20]. Therefore, there is a requirement for an automatic pest identification system.

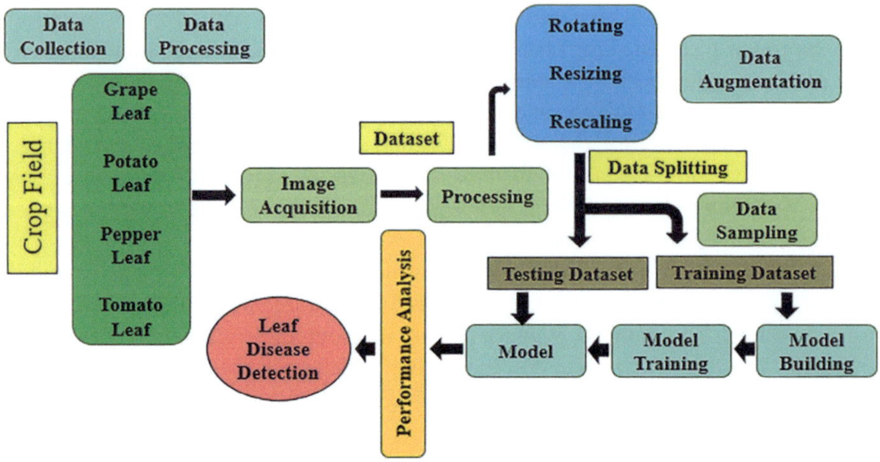

Fig. 4 Crop health monitoring using AI and MI technology

Artificial intelligence and machine learning play a fundamental role in the detection of pests in agriculture. These technologies leverage sophisticated algorithms and data analysis to effectively detect and diagnose a range of crop-related issues, facilitating the precise application of harmful chemicals based on specific needs. [21]. Cutting-edge pest detection systems, powered by real-time artificial intelligence and machine learning technologies, can precisely detect pests and diseases in crops by analysing images. Through the utilization of these sophisticated systems, automated pesticide sprayers can be triggered to target only the affected areas, leading to a more accurate pesticide application and a decrease in wastage. Figure 4, illustrates a detailed step-by-step explanation of the procedure for identifying diseases in plants.

Intelligent automation algorithms can recognise the requirement for pesticides promptly and provide accurate applications by focusing on particular regions. In case of an intrusion, instant notifications are dispatched to farmers' mobile devices, enabling them to act accordingly. Numerous authors have utilized artificial intelligence, machine learning, and cutting-edge techniques to detect pests and diseases, as illustrated in Table 3. The latest AI-powered drone technology, integrated with spraying mechanisms, can effectively cover extensive areas with a precision level of 90%, enhancing resource efficiency, cutting down expenses, and decreasing waste by as much as 80% through targeted applications, leading to substantial advancements in agricultural pest control [22]. As AI and ML continue to advance, their part in agriculture is projected to expand, offering even greater potential for enhancing food safety and agricultural sustainability.

Table 3 Pest and disease identification using AI and ML

S. No.	Intended work pre-specified	Methodology	Accuracy (%)	References
1	Identification of plant diseases and pests in eight types of horticultural crops	The ResNet-50 model with additional support of Vector Machine (SVM) classifier was used to identify pests and diseases in horticultural crops	98	[23]
2	Detecting pests and diseases on apple leaves	AlexNet precursor set was used for detecting pests and diseases in agriculture	97.62	[24]
3	Pests and diseases affecting apple fruits	The disease detection system based on fuzzy rules (DDSF) uses fuzzy logic to accurately identify diseases, improving accuracy and efficiency	91.66	[25]
4	Detection of insects and pests for pre-defined different types of crop	Utilized advanced deep learning methods such as faster R-CNNs, SSDs, and Yolo-v4 for image-based scale pest detection and localisation	89	[26]

2.4 Weed Management

Effective weed management is essential for successful crop production, but it can be challenging and time-consuming. Weeds compete with crops for resources and some are toxic, posing a threat to public health. While herbicide spray is commonly used, it can harm public health and cause environmental pollution if overused. In agriculture, invasive weeds present a major obstacle to productivity. Farmers face difficulties in manually identifying and removing each weed, resulting in heavy reliance on herbicides. However, herbicides contain harmful chemicals that can negatively impact crop and soil health, posing risks to human health as well. Unfortunately, over 90% of herbicides are misapplied, leading to environmental loss, failure to reach the intended target, and ineffective weed control in crops [27]. A study by Ronal Gerhards in University of Sheffield (UK) on using AI-driven herbicide spraying techniques can save more than 50% of herbicide in various crops without causing. The overuse of herbicides has caused herbicide-resistant weeds to become more prevalent. Manual weeding is labour-intensive and time-consuming, posing challenges for farmers who struggle to find enough workers in the agriculture sector, prompting the search for alternative methods.

AI and ML technologies can revolutionize weed management by providing more effective and precise techniques for identifying and regulating weeds. They are currently used for tasks such as weed identification, precise control, predictive modelling, and mapping. Different types of weed detection methods using AI technologies are shown in Fig. 5. Algorithmic intelligence can be trained to differentiate between crops and weeds using techniques such as CNN, DCNN, SVM, ANNs, RF classifier, KNN, ShuffleNet-v2, and VGGNet. Laser weeding technology provides chemical-free, no-till weed control for crops by identifying and eliminating

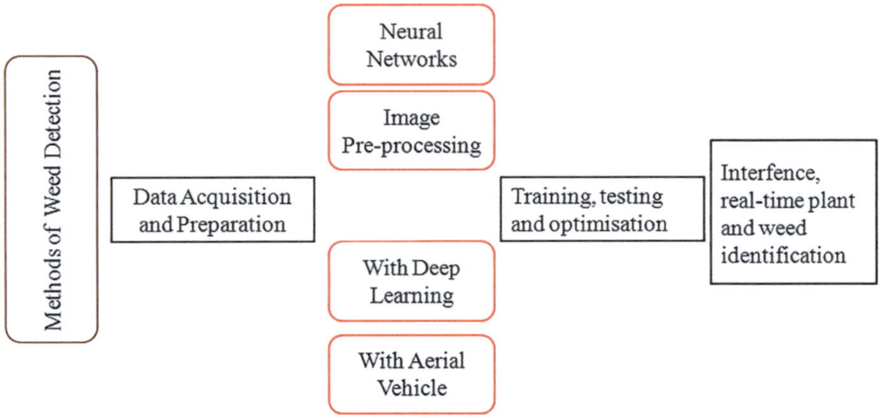

Fig. 5 Flowchart representation of weed identification methods

weeds early in their lifecycle, before they are visible to the human eye, preventing damage to crops. It offers precise targeting of weeds, including those between crops, with millimeter accuracy [28]. The integration of AI-ML in weed management systems ensures precision, cost-effectiveness, environmental sustainability, labour efficiency, increased crop yields, time efficiency, scalability, data-driven insights, climate resilience, and technological progress.

2.5 *Crop Quality*

Artificial intelligence sensors integrated with algorithmic learning play a fundamental role in enhancing crop quality as they deliver real-time data and analysis to farmers. These sensors are designed to monitor various aspects of crop growth, such as soil moisture levels, temperature, nutrient content soil quality, weather patterns, and crop diseases. By identifying patterns and correlations within the data, automated learning models can help agriculturalists make well-versed verdicts about irrigation, chemical fertilization application, and pest control, ultimately resulting in higher crop yields and better-quality produce [19]. These big data models can be used to foresee various crop issues in real-time, allowing farmers to take active measures to avoid or mitigate potential damage to their crops as shown in Table 4. By analysing old data and real-time environmental factors, these models can provide early warnings about potential threats to crop quality, enabling farmers to implement targeted interventions and minimize the impact on their harvest to ensure a higher yield of high-quality produce.

Furthermore, algorithmic models can improve the use of resources such as water for irrigation and agrochemicals, ensuing to more sustainable and environmentally friendly agricultural practices. By accurately predicting the water and nutrient needs

Table 4 Crop health monitoring using AI and ML-developed models and algorithms

S. No.	Categories specified in the assignment	Model/Techniques/Methods used	Accuracy/Results	References
1	Diseased and health status detection in 12 different species and 42 different classes	AlexNet, VGG 19, inception, DenseNet, ResNet, plant DiseaseNet object detection: two-stage methods—Faster R-CNN, faster R-CNN with TDM, faster R-CNN with FPN, one-stage methods—YOLOv3, SSD513, retina net	94%	[30]
2	Wheat (diseased and healthy)	Deep CNN model	Accuracy more than 96%	[31]
3	Maize (diseased and healthy)	Custom CNN model	92.85% accuracy	[32]
4	Rice (diseased and healthy)	Pre-trained VGGNet	Accuracy:91.83%	[33]

of crops, these models can support reduce waste and abate the environmental impact of farming operations, while still maintaining high crop quality. Overall, machine learning models are irreplaceable tools for modern agriculture, offering insights and forecasts that can considerably improve crop quality and yield. In addition to monitoring crop health, artificial intelligence sensors can also optimize harvesting processes to further improve crop quality. By analysing data on factors like ripeness and sugar content, farmers can determine the optimal time to harvest each crop, ensuring that it is picked at peak freshness and flavour [29]. This precision harvesting not only improves the quality of the crop but also reduces waste and maximizes profitability for farmers. Ultimately, these technologies recommend favourable solutions to these issues by identify patterns and trends that help farmers to take well-versed decisions to increase the eminence of their crops.

2.6 Plant Phenotyping

Studying plant phenotyping is essential for understanding plant-environment interactions, especially in crop management and breeding. Conventional phenotyping methods, involving manual measurements and observations, are slow, laborious, and liable to errors. The use of artificial intelligence and machine learning has transformed phenotyping by empowering the scrutiny of vast datasets and uncovering hidden patterns beyond human perception. High throughput imaging techniques for non-destructive phenotypic measurement are becoming increasingly popular. This system produces a large volume of images for quick and accurate phenotypic analysis, facilitating diverse phenomics studies. The assimilation of high-throughput imaging systems with advanced AI technologies, enhances both the effectiveness and accuracy of this field [34]. Phenomics has been applied to investigate various phenotypic traits, including spike detection and counting, yield prediction, assessment of plant senescence, leaf weight and count, plant volume, convex hull analysis, water stress evaluation, and numerous other aspects, as detailed in Table 5.

This is especially beneficial in the area of plant phenotyping, where artificial intelligence can recognize leaf shapes, assess plant growth parameters, and identify disease symptoms [1]. The field of high-phenotyping often involves several types of data, including imaging, genomic, and environmental data. This all-inclusive approach permits for a more thorough understanding of plant characteristics and behaviour, leading to advancements in agricultural research and crop improvement. Advanced fusion techniques, such as deep multimodal learning and graph-based models, enable researchers to uncover hidden patterns and connections among different data sources. Furthermore, real-time phenotyping allows the continuous monitoring of plant traits throughout the growth cycle, providing valued understandings into dynamic responses to environmental conditions. It will enable rapid decision-making and adaptive management strategies in agriculture by optimizing resource allocation and improving yield and quality [35].

Table 5 Phenotyping using AI and ML-developed models and algorithms

S. No.	Categories specified in the assignment	Model/ Techniques/ Methods used	Accuracy/Results	References
1	Spike recognition	Neural networks and texture energy laws were used in this study	80%	[36]
2	Spike recognition in the field	Faster R-CNN	88–94%	[37]
3	Spike recognition and count	U-Nets	99.93%,	[38]
4	NDVI	RGB image manipulation	More cost-effective and user-friendly than traditional dual image NDVI or hyper-spectral imaging methods	[39]

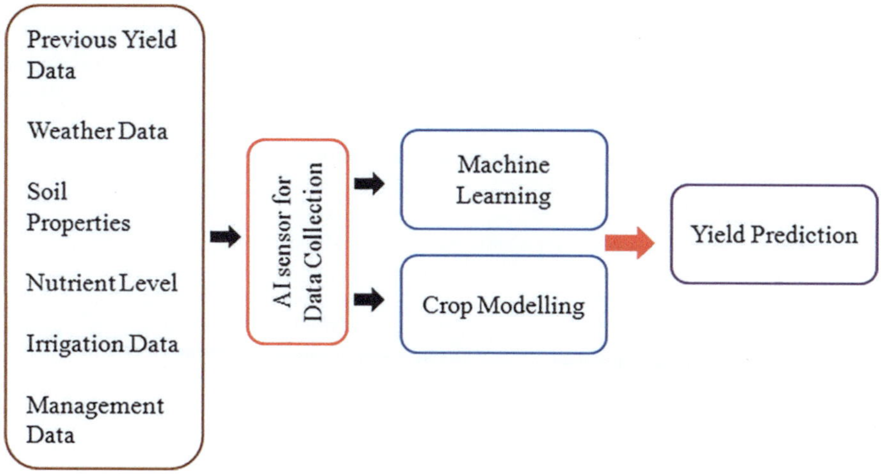

Fig. 6 The framework of yield prediction models

2.7 *Yield Prediction*

Artificial intelligence sensors play a major role in predicting crop yields. These sensors are deliberate to collect and analyse data related to various environmental aspects such as temperature, humidity, soil moisture, and light intensity. By constantly monitoring these parameters, AI sensors can deliver valued perceptions into the growth and important real information [40]. This statistics is then used to sort accurate predictions about the potential yield of a particular crop by developing predictive models by training the algorithmic learning models with massive datasets, allows for more accurate and dependable predictions (Fig. 6) and Table 6. This AI-ML driven predictive capability allows farmers to make cognisant verdicts regarding planting and irrigation schedules, fertilization, and pest control measures to adjust crop production.

2.8 *Livestock Management*

The livestock industry in India made up around 4.11% of the GDP and 25.6% of the agricultural GDP in 2023, as reported by the Ministry of Fisheries, Animal Husbandry & Dairying. Despite its consistent growth, the sector continues to rely on manual control and monitoring, with alternative technologies being uncomfortable, stressful, or costly. As a result, there is a noteworthy need for modern technological interventions in this sector. AI and ML are increasingly being utilized in the arena of livestock management. These technologies offer a wide range of benefits, including improved efficiency, accuracy, and productivity [44].

Table 6 Remote sensing and machine learning approaches for yield estimation in diverse crops

S. No.	Categories specified in the assignment	Model/ Techniques/Methods used	References
1	Research is using vegetation indices and machine learning with remote sensing technology to predict crop yields in the Canadian Prairies	Various techniques such as multiple linear regression and neural networks were utilized for data analysis	[41]
2	Predicting wheat yields in the south eastern region of Turkey through the application of artificial neural networks	Neural networks and multivariate polynomial regression computational techniques were used for modelling multifarious relationships in data	[42]
3	A method utilizing artificial intelligence has been developed to forecast the Robusta coffee yield based on soil fertility characteristics	ELM, random forest, and multiple linear regression were used in the study	[43]

AI and ML algorithms are employed to oversee the well-being and health of livestock. These sensors can identify variations in body temperature, activity level, and other physiological parameters. Precision livestock farming utilizes non-invasive sensors like cameras, accelerometers, gyroscopes, radio-frequency identification systems (RFID systems), pedometers, and optical and temperature sensors. IoT sensors detect variable physical quantities (VPQs) to monitor temperature, sound, humidity, etc. These sensors can alert farmers in real time if a VPQ deviates from normal levels, providing crucial insights into each animal. Checking each animal repeatedly and laboriously can be made more cost-effective [45].

3 Cost Implications of Integrating AI and ML into Agriculture

The financial considerations associated with incorporating AI and ML into the agricultural sector is significant and multifaceted. One of the primary cost implications is the initial investment required to implement AI-ML technologies, such as purchasing the necessary hardware and software, along with training personnel to use and maintain these systems. Additionally, ongoing costs related to data storage, software updates, and technical support need be taken into consideration when budgeting for AI integration in agriculture. Another cost apprehension is the potential impact on labour expenses. While AI has the power to streamline operations and increase efficiency, it may also lead to a reduction in the need for human labour in certain tasks. This could result in cost savings for some agricultural operations, but it may also require retraining or reallocating workers to other roles within the organization, which can incur additional expenses [46].

Furthermore, it is indispensable to consider the long-term financial benefits of integrating AI and ML into agriculture, such as increased productivity, improved crop yields, and enhanced decision-making capabilities. While the upfront costs of AI and ML implementation may be significant, the potential return on investment in terms of increased efficiency and profitability can outweigh these initial expenses in the long run. Ultimately, careful financial planning and strategic decision-making are essential when considering the cost implications of integrating AI-ML into agriculture.

4 Challenges Associated with AI and ML Technologies to Small-Scale Farmers

AI and ML have the potential to significantly impact the agricultural segment, particularly in addressing the problems faced by small-scale farmers. By utilizing AI and ML technologies, farmers can access appreciated understandings and data-driven solutions to increase crop yields, optimize resource management, and enhance overall productivity. These technologies are able to analyse massive amounts of data, such as weather patterns, soil health, and market inclinations, to provide personalized recommendations and strategies for farmers to make conversant verdicts [47].

One of the key benefits of AI and ML in agriculture is their capability to help small-scale farmers overcome various challenges they encounter daily. For instance, AI-powered tools can backing farmers in predicting crop diseases and other potential risks, allowing them to take pre-emptive measures to protect their crops and maximize their harvest. Additionally, ML algorithms can help farmers improve irrigation schedules, fertilizer usage, and pest control methods, leading to more sustainable farming practices and increased profitability.

5 Training or Skills are Required for Farmers to Adopt AI Technology

To effectively implement AI and ML technology in farming, farmers need to acquire a certain set of training and skills. Firstly, they should have a strong understanding of the principles and concepts behind artificial intelligence and machine learning. This comprises knowledge of algorithms, data analysis, and programming languages commonly used in AI and ML applications. Additionally, farmers should be proficient in utilizing AI and ML tools and platforms specific to the agricultural industry, such as precision agriculture software and autonomous farming equipment. Furthermore, farmers need to improve expertise in data collection and management, as AI and ML technology heavily rely on large datasets for analysis and decision-making [48]. This involves knowing how to gather and process various types of agricultural data, such as crop yields, soil quality, weather patterns, and pest infestations.

Moreover, farmers should be adept at interpreting the insights generated by AI and ML models to sort well-versed conclusions about crop management, resource allocation, and overall farm operations. In addition to technical knowledge and data skills, farmers should also possess a mindset open to innovation and continuous learning. Embracing AI and ML technology requires a willingness to adapt to new methods and tools, as well as a proactive approach to staying updated on the most recent advancements in agricultural technology. Moreover, effective communication and collaboration with agritech experts, data scientists, and AI specialists can further enhance a farmer's ability to successfully adopt and integrate AI and ML technology into their farming practices.

6 Limitations and Challenges of AI and ML in Agriculture

The constraints and obstacles faced by AI and ML in the field of agriculture are multifaceted. One limitation is the lack of high-quality data required for training AI and ML algorithms. Agriculture involves a wide range of variables such as soil quality, weather conditions, and crop health that makes it challenging to collect and analyse comprehensive datasets. Insufficient data can hinder AI-ML models from providing accurate predictions or recommendations for farmers.

The complexity of agricultural systems presents another challenge, as these systems can vary significantly based on factors such as geographical location, type of crops, and farming methods. Developing AI-ML solutions that are flexible to different agricultural contexts can be challenging due to this variability [49]. Additionally, the implementation of AI-ML technologies in agriculture requires significant investment in infrastructure, training, and maintenance, which may be a barrier for small-scale farmers or developing countries with limited resources.

Furthermore, ethical considerations such as data privacy, algorithm bias, and job displacement also pose challenges for the adoption of AI-ML in agriculture. Farmers might be reluctant to disclose confidential information regarding their operations to AI systems and there is a risk of unintended consequences if algorithms are not designed and monitored carefully [50]. Addressing these limitations and challenges will have need of collaboration between researchers, policymakers, and industry stakeholders to safeguard that these technologies are effectively integrated into agricultural practices while minimizing potential risks.

7 Conclusion

The integration of AI-ML technologies has the power to offer effective solutions to significant challenges in agriculture, including soil health monitoring, irrigation planning, crop disease management, pest identification, crop phonemics, and more. By leveraging AI and ML, farmers can sort data-driven decisions that regulate crop yields and enhance overall agricultural productivity. The implementation of AI-related solutions in the agriculture domain is expected to drive innovation, increase efficiency, and ultimately contribute to the sustainable growth of the agricultural sector. Future studies on incorporating ML in agriculture should prioritize utilizing various data sources, including satellite/drone imagery, IoT-based sensor data, and weather station information, to gain a deeper insight into agricultural systems. Furthermore, combining ML with robotics and automation offers the prospective for intelligent, self-learning systems that can convey intricate tasks for farmers and agricultural activities, such as the creation of autonomous fruit-picking machines.

The primary objective moving forward should be the development of cost-effective and adaptable machine-learning solutions tailored for regions with constrained resources, with a particular emphasis on extending the advantages of this technology to small-scale farmers and communities in emerging economies. By pursuing these research avenues, advancements can be made in establishing more enduring, productive, and robust agricultural frameworks. Collaborations of subject experts and professionals in agricultural engineering, agronomy, or soil science can result in customized solutions for agricultural issues. Additionally, conducting comprehensive research to evaluate the socio-economic consequences of the extensive usage of machine learning in agriculture, including its impact on employment, economic sustainability, and fair access to technological resources, would be a valuable endeavour.

Acknowledgements The authors express their gratitude for the technical and moral assistance received from the Faculty of Agricultural Engineering, SKUAST-Jammu, as well as for the practical training in Sensor Based Smart Agriculture (Project No: 17) under the Holistic Agricultural Development Program in Jammu and Kashmir.

Author Contribution All authors have equal contribution.

Data Availability There is no data available.

Statements and Declarations Authors declare that there is no significant competing financial, professional, or personal interests that might have influenced the performance or presentation of the work described in this manuscript.

References

1. Mathur R (2023) Artificial intelligence in sustainable agriculture. Int J Res Appl Sci & Eng Technol 11(6):4047–4052. https://doi.org/10.22214/ijraset.2023.54360
2. Linaza MT, Posada J, Bund J, Eisert P, Quartulli M, Döllner J, Lucat L (2021) Data-driven artificial intelligence applications for sustainable precision agriculture. Agronomy 11(6):1227. https://doi.org/10.3390/agronomy11061227
3. Leong YM, Lim EH, Subri NF, Jalil N (2023) Transforming agriculture: navigating the challenges and embracing the opportunities of artificial intelligence of things. In: 2023 IEEE international conference on agrosystem engineering, technology & applications (AGRETA). Shah Alam, Malaysia, pp 142–147. https://doi.org/10.1109/AGRETA57740.2023.10262747

4. Larson DB, Harvey H, Rubin DL, Irani N, Justin RT, Langlotz CP (2021) Regulatory frameworks for development and evaluation of artificial intelligence-based diagnostic imaging algorithms: summary and recommendations. J Am Coll Radiol 18(3):413–424. https://doi.org/10.1016/j.jacr.2020.09.060
5. Kulykovets O (2023) Automation of production processes in agriculture using selected artificial intelligence tools. Ann Pol Assoc Agric Agribusiness Econ 25(4):255–267. https://doi.org/10.5604/01.3001.0053.9616
6. Javaid M, Haleem A, Khan IH, Suman R (2023) Understanding the potential applications of artificial intelligence in agriculture sector. Adv Agrochem 2(1):15–30. https://doi.org/10.1016/j.aac.2022.10.001
7. Ikrang EG, Unwana IU, Precious OE (2022) The use of artificial intelligence in tractor field operations: a review. Poljoprivredna tehnika 47(4):1–14. https://doi.org/10.5937/poljteh2204001g
8. Helm JM, Swiergosz AM, Haeberle HS, Karnuta JM, Schaffer JL, Krebs VE, Spitzer AI, Ramkumar PN (2020) Machine learning and artificial intelligence: definitions, applications, and future directions. Curr Rev Musculoskelet Med 13:69–76. https://doi.org/10.1007/s12178-020-09600-8
9. Gardezi M, Joshi B, Rizzo DM, Ryan M, Prutzer E, Brugler S, Dadkhah A (2023) Artificial intelligence in farming: challenges and opportunities for building trust. Agron J 116(3):1217–1228. https://doi.org/10.1002/agj2.21353
10. Gupta N, Gupta P, Nadeem D, Abuzar A, Elahi A (2020) Artificial intelligence in agriculture. J Phys: Conf Ser 1693(1):012058. https://doi.org/10.1088/1742-6596/1693/1/012058
11. El-Bilali A, Taleb A, Brouziyne Y (2021) Groundwater quality forecasting using machine learning algorithms for irrigation purposes. Agric Water Manag 245:106625. https://doi.org/10.1016/j.agwat.2020.106625
12. Vangala A, Kumar DA, Chamola V, Korotaev V, Rodrigues JPPC (2023) Security in IoT-enabled smart agriculture: Architecture, security solutions and challenges. Clust Comput 26(2):879–902
13. Neupane J, Guo W (2019) Agronomic basis and strategies for precision water management: a review. Agronomy 9:87. https://doi.org/10.3390/agronomy9020087
14. Suchithra MS, Pai ML (2019) Improving the prediction accuracy of soil nutrient classification by optimizing extreme learning machine parameters. Inf Process Agric 7(1):72–82
15. Yang M, Xu D, Chen S, Li H, Shi Z (2019) Evaluation of machine learning approaches to predict soil organic matter and pH using vis-NIR spectra. Sensors (Switzerland) 19(2):263–277
16. Morellos A, Pantazi X, Moshou D, Alexandridis T, Whetton R, Tziotzios G, Wiebensohn J, Bill R, Mouazen AM (2016) Machine learning based prediction of soil total nitrogen, organic carbon and moisture content by using VIS-NIR spectroscopy. Biosyst Eng 152:104–116. https://doi.org/10.1016/j.biosystemseng
17. Ramya S, Swetha AM, Doraipandian M (2020) IoT framework for smart irrigation using machine learning technique. J Comput Sci 16:355–363
18. Ranjan R, Sahoo RN, Chopra UK, Pramanik M, Singh AK, Pradhan S (2017) Assessment of water status in wheat (Triticum aestivum L.) using ground based hyperspectral reflectance. Proc Natl Acad Sci India Sect B Biol Sci 87(2):377–388
19. Fuentes AF, Yoon S, Lee J, Park DS (2018) High-performance deep neural network-based tomato plant diseases and pests diagnosis system with refinement filter bank. Front Plant Sci 9:1162. https://doi.org/10.3389/fpls.2018.01162
20. Mokaya V (2019) Future of precision agriculture in India using machine learning and artificial intelligence. Int J Comput Sci Eng 7(2):1020–1023
21. Pineda M, Pérez-Bueno ML, Barón M (2018) Detection of bacterial infection in melon plants by classification methods based on imaging data. Front Plant Sci 9:164. https://doi.org/10.3389/fpls.2018.00164
22. Zhang X, Qiao Y, Meng F, Fan C, Zhang M (2018) Identification of maize leaf diseases using improved deep convolutional neural networks. IEEE Access 6:30370–30377. https://doi.org/10.1109/ACCESS.2018.2844405

23. Turkoglu M, Hanbay D (2019) Plant disease and pest detection using deep learning-based features. Turk J Electr Eng Comput Sci 27:1636–1651. https://doi.org/10.3906/elk-1809-181
24. Liu B, Zhang Y, He D, Li Y (2018) Identification of apple leaf diseases based on deep convolutional neural networks. Symmetry 10(1):11. https://doi.org/10.3390/sym10010011
25. Kour V, Arora S (2019) Fruit disease detection using rule-based classification. In: Proceedings of smart innovations in communication and computational sciences, advances in intelligent systems and computing (ICSICCS-2018), pp 295–312
26. Cheeti S, Kumar GS, Priyanka JS, Firdous G, Ranjeeva PR (2021) Pest detection and classification using YOLO AND CNN. Ann Roman Soc Cell Biol 25:15295–15300
27. Su J, Zhu X, Li S, Chen WH (2023) AI meets UAVS: a survey on AI AI-empowered UAV perception systems for precision agriculture. Neurocomputing 518:242–270
28. Wang A, Zhang W, Wei X (2019) A review on weed detection using ground-based machine vision and image processing techniques. Comput Electron Agric 158:226–240. https://doi.org/10.1016/j.compag.2019.02.005
29. Barbedo JGA (2019) Plant disease identification from individual lessons and spots using deep learning. Biosyst Eng 180:96–107. https://doi.org/10.1016/j.biosystemseng.2019.02.002
30. Acemoglu D, Restrepo P (2019) Artificial intelligence, automation, and work. In: The economics of artificial intelligence, pp 197–236. https://doi.org/10.7208/chicGO/9780226713475.003.0008
31. Picon A, Alvarez-Gila A, Seitz M, Ortiz-Barredo A, Echazarra J, Johannes A (2019) Deep convolutional neural networks for mobile capture device-based crop disease classification in the wild. Comput Electron Agric 161:280–290
32. Sibiya M, Sumbwanyambe MA (2019) Computational procedure for the recognition and classification of maize leaf diseases out of healthy leaves using convolutional neural networks. Agric Eng 1(1):119–131
33. Chen J, Zhang D, Nanehkaran YA, Li D (2020) Detection of rice plant diseases based on deep transfer learning. J Sci Food and Agric 100(7):3246–3256
34. Kumar V, Masrat M (2024) Artificial intelligence robotics technologies for harvesting horticultural crops: an alternative management approach. Indian J Ecol 51(6):1585–1595. https://doi.org/10.55362/IJE/2024/4445
35. Ampatzidis Y (2018) Applications of artificial intelligence for precision agriculture. Edis 1–5. https://doi.org/10.32473/edis-ae529-2018
36. Qiongyan L, Cai J, Berger B, Miklavcic S (2014) Study on spike detection of cereal plants. In: 13 international conference on control automation robotics & vision (ICARCV). IEEE, pp 228–233
37. Hasan MM, Chopin JP, Laga H, Miklavcic SJ (2018) Detection and analysis of wheat spikes using convolutional neural networks. Plant Methods 14(1):1–13
38. Misra T, Arora A, Marwaha S, Chinnusamy V, Rao AR, Jain R (2020) Spike SegNet-a deep learning approach utilizing encoder-decoder network with hourglass for spike segmentation and counting in wheat plant from visual imaging. Plant Methods 16(1):1–20
39. Beisel NS, Callaham JB, Sng NJ, Taylor DJ, Paul A, Ferl RJ (2018) Utilization of single-image normalized difference vegetation index (SI-NDVI) for early plant stress detection. Appl Plant Sci 6(10):3–10
40. Anwarul S, Misra T, Srivastava D (2022) An IoT & AI-assisted framework for agriculture automation. In: 10th international conference on reliability, Infocom technologies and optimization (trends and future directions) (ICRITO). Noida, India, pp 1–6. https://doi.org/10.1109/ICRITO56286.2022.9964567
41. Johnson MD (2013) Crop yield forecasting on the Canadian prairies by satellite data and machine learning methods. Master's thesis, University of British Columbia, Atmospheric Science
42. Cakir Y, Kirci M, Gunes EO (2014) Yield prediction of wheat in south-east region of Turkey by using artificial neural networks. In: 2014 the 3rd international conference on agro-geoinformatics, agro-geoinformatics (2014). https://doi.org/10.1109/Agro-Geoinformatics.2014.6910609

43. Kouadio L, Deo RC, Byrareddy V, Adamowski JF, Mushtaq S, Phuong NV (2018) Artificial intelligence approach for the prediction of Robusta coffee yield using soil fertility properties. Comput Electron Agric 155:324–338. https://doi.org/10.1016/J.COMPAG.2018.10.014
44. Akhigbe BI, Munir K, Akinade O, Akanbi L, Oyedele LO (2021) IoT technologies for livestock management: a review of present status, opportunities, and future trends. Big Data Cogn Comput 5:10. https://doi.org/10.3390/bdcc5010010
45. Binch A, Fox CW (2017) Controlled comparison of machine vision algorithms for Rumex and Urtica detection in grassland. Comput Electron Agric 140:123–138
46. Bao J, Xie Q (2022) Artificial intelligence in animal farming: a systematic literature review. J Clean Prod 331:129956. https://doi.org/10.1016/j.jclepro.2021.129956
47. Bokonda PL, Ouazzani-Touhami K, Souissi N (2020) Predictive analysis using machine learning: review of trends and methods. In: 2020 international symposium on advanced electrical and communication technologies, ISAECT (2020). https://doi.org/10.1109/ISAECT50560.2020.9523703
48. Fuentes S, Gonzalez Viejo C, Tongson E, Dunshea FR (2022) The livestock farming digital transformation: implementation of new and emerging technologies using artificial intelligence. Anim Health Res Rev 23(1):59–71
49. Helwatkar A, Riordan D, Walsh J (2014) Sensor technology for animal health monitoring. Int J Smart Sens Intell Syst 7(5):1–6. https://doi.org/10.21307/IJSSIS-2019-057
50. Lakshmi V, Corbett J (2020) How artificial intelligence improves agricultural productivity and sustainability: a global thematic analysis. In: Proceedings of the 53rd Hawaii international conference on system sciences, pp 5202–5211. http://hdl.handle.net/10125/64381

Evolution of Object Detection: From Classical to Modern AI Approaches

Pankaj Verma, Gaurav Verma, Chhagan Charan, and Jyoti Ramola

Abstract Agile and correct detection of objects is becoming more and more crucial with the increased interest in the area of smart video surveillance, autonomous vehicles, facial recognition, and various distinct applications. These systems not only identify and categorize objects within images and videos but also precisely locate them by describing bounding boxes. This paper administers a detailed analysis of traditional and modern deep learning-based approaches for detection of objects, examining aspects such as multi-scale feature recognition, data augmentation techniques, training methodologies, and viewpoint variability. Key standard datasets utilized in object detection research are also reviewed. Additionally, the paper discusses current challenges and outlines future research directions, particularly focusing on evolving datasets and frameworks that underpin object detection tasks. The analysis reveals that while existing object detection methods perform reasonably well, there remains significant room for improvement, especially in scenarios involving large variations in object scales, occluded views, and challenging environmental conditions. Consequently, the paper suggests avenues for advancement in object detection techniques.

Keywords Object detection · Computer vision · Deep learning · Machine learning

P. Verma (✉) · G. Verma · C. Charan
ECE Department, National Institute of Technology, Kurukshetra, Haryana, India
e-mail: pankaj@nitkkr.ac.in

G. Verma
e-mail: gaurav@nitkkr.ac.in

C. Charan
e-mail: chhagan.charan@nitkkr.ac.in

J. Ramola
ECE Department, Graphic Era Hill University, Dehradun, India
e-mail: jramola777@gmail.com

S. Chand Kumain et al. (eds.), *AI and ML Techniques in Image Processing and Object Detection*, https://doi.org/10.1007/978-981-96-7445-9_3

1 Introduction

Computer vision is an intriguing field of study which enables computers to extract, analyze and intercept the information from images or video data, in a manner similar to human beings. Object detection is a subfield of computer vision which aims to recognize and isolate different articles in images and videos, and has wide variety of applications like autonomous vehicles [1–3], face recognition [4–6], video surveillance [7, 8], face detection [9, 10], traditional object detection [11–14] etc.

The problem of object detection can be stated as localizing and identify different types of objects in images. The evolution of object detection is primarily classified into two categories: the methods uses prior to 2014 known as classical approaches and the methods which are based on deep learning known as Modern AI approaches. In this paper, we will start with the review of traditional approaches and then move on to the state of the art processes which are established on modern artificial intelligence techniques. The primary outputs of the manuscript can be listed as:

a. Basic discussion of the traditional approaches of object detection
b. Critical review of the modern deep learning approaches for objection detection and their characteristics. Main datasets and performance metrics are also discussed.
c. Future directions for improving the objection detection

The remaining manuscript is arranged as follows: Sect. 2 covers the traditional approaches of detection of objects, Sect. 3 describes the artificial intelligence based approaches for object detection, which covers the different types of single stage and two stage detectors. Data sets and various performance metrics used in object detection approaches has been discussed in Sect. 4. The work which can be carried out in future is described in Sect. 5 and finally the paper is concluded in Sect. 6.

2 Traditional Approaches

In the era of traditional object detection, maximum number of the methods were established on handcrafted characteristics because of the inadequacy of efficacious image depiction. Some of the techniques are explained as follows.

2.1 Viola Jones Detectors

Proposed in 2001 by Michael Jones and Paul Viola [15], this framework for object detection enables real-time detection of human faces. It employs a method of sliding windows across an image at various scales and positions to identify regions containing human faces. These sliding windows search for 'haar-like' features, named after Alfred Haar, who pioneered haar wavelets.

The framework utilizes haar wavelets to represent image characteristics. To enhance detection speed, it integrates an integral image, ensuring that the implementation complexity of each sliding window remains independent of its size. Additionally, the authors employed the Adaboost algorithm for characteristics selection, which identifies a subset of features crucial for detection of face among a large pool of random features. The framework also incorporates Detection Cascades, a multi-stage detection approach that reduces computational load by prioritizing face targets over background windows during processing, under the condition that the first paragraph of a section or subsection is not indented.

2.2 *HOG Detector*

Initially introduced in 2005 by Triggs and Dalal [16], Histogram of Oriented Gradients (HOG) represents an advancement over contemporary methods like Scale Invariant Feature Transform and Shape Contexts. This detector operates by dividing the image into blocks (similar to a sliding window) and employs a dense pixel grid where gradients are calculated based on changes in pixel intensities' magnitude and direction within each block.

HOG is notably recognized for its application in detection of pedestrians. For accommodating objects of varying sizes, this method resizes the input image repeatedly while maintaining the size of the detection window as constant.

2.3 *Deformable Part-Based Model (DPM)*

In 2008, this detector was initially proposed by Felzenszwalb et al. [17] as a development over the HOG detector, the Deformable Parts Model (DPM) has seen various enhancements by R. Girshick. The approach tackles the challenge of detecting complex objects like cars through a 'divide and conquer' strategy, distinguishing between the window, body, and wheels.

The training phase of DPM involves learning to decompose objects effectively, while inference combines detections from distinct portions of the object ensemble. The DPM detector consists of a root-filter and multiple part-filters. The weakly supervised learning strategy within DPM naturally learns configurations (such as size and location) of part filters as latent variables.

To enhance detection precision, R. Girshick introduced techniques like a specialized form of Multi-Instance learning, "hard negative mining," "bounding box regression," and "context priming." Additionally, authors implemented a cascade architecture that significantly accelerates processing speed by more than tenfold without compromising accuracy.

3 Modern Deep Learning Based Approaches

Object detection faced a stagnation after 2010 due to the limitations of hand-crafted features reaching their performance ceiling. However, a significant breakthrough occurred in 2012 amidst of the resurgence of convolutional neural networks (CNNs), which proved highly effective at learning complex and relevant feature representations from images. This led to a revitalization of object detection techniques. The deadlock in object detection was decisively broken in 2014 by the addition of Regions with CNN features (RCNN) [18]. In the current era dominated by deep learning, object detection methodologies are primarily categorized into two main approaches: "two-stage detection" and "one-stage detection". Table 1 summarizes the evolution of various one stage and stage detectors over the years.

3.1 One Stage Detectors

One-stage detectors like SSD (Single Shot MultiBox Detector) and YOLO (You Only Look Once) have become popular because of their straightforwardness and potential to operate in real-time. They estimates object bounding boxes and the class probabilities straight in lone sweep over the image, thus eliminating the requirement for a separate stage for region proposals. One stage detectors are simpler, faster and are robust to scale changes. However, their accuracy is lower and low robust to occlusions.

3.1.1 Over Feat

This method first introduced a comprehensive framework using Convolutional Networks that integrates classification, localization, and detection through a multi-scale sliding window method [19]. This integrated framework was a conquerer in the ImageNet Large Scale Visual Recognition Challenge 2013 (ILSVRC) in the localization tasks and attained competing results in classification and detection. It is a single-stage detector which performs object detection in a sole progressing pass

Table 1 Evolution of one stage and two stage detectors over the years

Sr. No.	One stage detector	Year	Two stage detectors	Year
1	Over feat	2013	RCNN	2014
2	Retina net	2013	SPP Net	2014
3	Single shot detector	2016	Fast RCNN	2015
4	Yolo	2016	Faster RCNN	2015
5	Yolo V2, V3 V4 and V5	2017 onwards	Mask RCNN	2017

using fully connected convolutional layers. This model has served as the foundation for subsequent techniques such as YOLO and SSD. A key distinction of OverFeat is its sequential training approach for classifiers and regressors.

3.1.2 Retina Net

RetinaNet, introduced in [20, 21], represents a one-stage detector that incorporates focal loss and FPN (Feature Pyramid Network) as key components. One-stage detectors often face challenges like lesser accuracy as a result of class imbalance. The FPN method addresses this issue by generating feature maps at multiple scales through lateral and top-down links across multiple levels. The focal loss function further enhances Retina Net's performance by adjusting the weights assigned to easily and hard-to-classify samples. This approach allows the network to effectively prioritize challenging samples, thereby balancing the treatment of imbalanced data. As a result, Retina Net achieves improved detection performance without compromising on the agility of one-stage detectors.

The architecture of Retina Net includes both a bottom-up pathway for feature extraction and a top-down pathway for integrating features across various layers using lateral connections. This design makes Retina Net a typical example of an FPN-based multi-scale detector, able of effectively handling objects at distinct scales within images.

3.1.3 SSD

The Single Shot Detector (SSD) was first introduced in 2016 [22]. This represents a single-stage model designed for multi-category prediction, co-existing with the YOLO series. The SSD method employs already defined set of anchor boxes that encompass various scales and aspect ratios to distinguish the bounding box outputs. This model integrates predictions from a number of feature maps at dissimilar resolutions, effectively addressing the problem of detecting objects with diverse ratios and scales.

SSD builds upon the VGG16 architecture by appending additional convolutional feature layers at the network's end to enable detection across multiple scales. During training, the network optimizes using a combined loss function comprising confidence and localization losses, weighted appropriately. Post-processing of detection outputs involves Non-Maximum Suppression (NMS) to consolidate the final outcomes. The SSD model incorporates the VGG16 convolutional network as its backbone, serving as a feature extractor through fully interconnected convolutional layers that streamlines the feature mapping phenomenon. Appended feature layers were introduced specifically to notice the broader features from input layers, thereby improving the model's capability for the detection of objects across various scales.

3.1.4 YOLO

The YOLO (You Only Look Once) model was initially developed by Redmon et al. [23]. This is a one-stage detector which treats detection of object as a regression problem. This method estimates the coordinates of bounding box for objects and assigns probabilities to determine their associated categories. By utilizing a single neural network, YOLO achieves end-to-end optimization. Unlike region-based methods that focus on specific regions for feature extraction, YOLO leverages features out of the entire image [24].

In this technique, an image is partitioned into a grid of size S $\times$ S. Every grid cell forecast five parameters: w, h (width and height of the bounding box), x, y (coordinates of the bounding box center), and a confidence score suggesting the existence of an object. The confidence score is based on the probability that an object is existing in the bounding box. This model assigns this confidence score to each class, and the class having the maximum probability is considered as the assigned class. The height (h) and width (w) parameters of the bounding box in YOLO are calculated as per the size of the object. During post-processing, overlapping bounding boxes are evaluated using Intersection over Union (IOU). The box having maximum IOU score is considered as the most accurate prediction for that object, while redundant or less accurate boxes are discarded.

3.1.5 YOLOv2

YOLOv2 [25], an improved iteration of YOLOv1 [23], was developed by Redmon, J. et al., designed to detect objects in images and videos with agility and accuracy. The accuracy is improved by using a more sophisticated network architecture and introducing techniques like batch normalization, which helps stabilize and accelerate training. It also incorporates Anchor boxes (predefined bounding boxes) to enhance the model's capability to detect objects of distinct sizes. It employs Darknet-19 as the backbone, consisting of more convolutional layers as compared to the Darknet architecture used in YOLO, that helps in reducing the processing required to study an image and at the same time obtaining higher accuracy.

3.1.6 YOLOv3

YOLOv3 [26] introduced "incremental improvements" over its predecessors. The authors replaced the Darknet-19 with a bulkier network architecture Darknet-53 and integrated features such as batch normalization, data augmentation and multi-scale training. Additionally, they replaced the softmax classifier layer with a logistical classifier. The speed of YOLOv3 was higher than YOLOv2 but it did not bring any groundbreaking changes and had lower accuracy in comparison to the cutting edge detectors from the previous year.

3.1.7 YOLOv4

YOLOv4 [27] incorporated some ingenious intelligence to create a fast object detector that can be trained with ease for existing systems. This version employs a "bag of freebies," that includes ways to increase training time without any change in the inference time, such as regularization methods, data augmentation techniques, class label smoothing, CIoU-loss, Cross mini-Batch Normalization (CmBN), self-adversarial training, and a cosine annealing scheduler. Additionally, this version has a "bag of specials" which includes features like DropBlock that can be turned on or off based on the specific use case. It also uses a genetic algorithm for hyper-parameter searching. YOLOv4 maintains the real-time speed characteristic of YOLO models while improving detection accuracy, making it suitable for applications where both high speed and high accuracy are critical.

3.1.8 YOLOv5

Just after the realization of YOLOv4, the Ultralytics company introduced the YOLOv5 repository [28, 29], which included significant improvements when compared to earlier versions of YOLO. It has been broadly adopted in numerous applications and has proven effective, enhancing the model's reliability. The inference speed of YOLOv5 was 140 fps and it uses PyTorch, making the implementation of the model easier, agile, and more accurate.

Despite the similarities between YOLOv4 and YOLOv5 architecture, YOLOv5 has demonstrated better performance than YOLOv4 in many scenarios. YOLOv5 comes in different sizes (s, m, l, x) to cater to different performance and computational needs.

In conclusion, the YOLOv5 model is an excellent choice for detecting small objects and is the quickest technique in comparison to other models. For real time applications, single stage detectors works fine, whereas for better accuracy, two stage detectors would be required. In the next section, we will go through the different types of two stage detectors.

3.2 Two Stage Detection

These type of detectors, exemplified by Faster R-CNN (Region-based Convolutional Neural Networks), have emerged in the form of leading methodologies for object detection. These detectors are structured around two pivotal stages: first is the proposal of the region and second is the classification of the object. In the initial stage, potential areas containing the objects are spotted through methods like Selective Search or Region Proposal Networks (RPNs). Following this, in the object classification stage, these proposed regions are utilized to class the objects and calibrate the bounding box predictions. These are more accurate, better localization possible and more robust to noise. However slower and more tedious as compared to one stage detectors.

3.2.1 RCNN

The Region Convolutional Neural Network (R-CNN) [30] was the beginning of this type of detectors which has proven how CNNs can significantly improve performance. It employs a class-agnostic module for proposal of region having CNNs to transform detection into a localization and classification problem. Initially, a mean-subtracted input image is processed with the help of the unit of region proposal. Thereafter, it identifies portion of the image with greater probability of containing an object by applying Selective Search [31]. The authors in [32] used AlexNet as the detector's backbone architecture. The feature vectors are subsequently fed into trained, class-specific Support Vector Machines (SVMs) to calculate confidence scores. Non-maximum suppression (NMS) is then used to the scored regions based on their class and IoU. After identifying the class, it predicts the bounding box with the help of a trained bounding-box regressor, that estimates the width, the height and the center coordinates of the box.

The training procedure of R-CNN is a bit complex. The initial step involves pre-training the CNN on a larger dataset. Next, the network is fine-tuned for detecting the objects by making use of domain-specific images (warped proposals, mean-subtracted) and substituting the classification layer by an arbitrarily initialized N + 1-way classifier, where N is the number of classes, employing stochastic gradient descent (SGD). It's training process was intricate, taking days for training on smaller datasets, even with shared computations.

3.2.2 SPP Net

In standard CNN architectures, pooling layers are often utilized to down sample feature maps, decreasing their dimensional parameters. This process typically involves operations like max pooling or average pooling. Further, conventional methods can lead to issues when dealing with images of different sizes or when the input image dimensions are not compatible with the fixed size expected by fully connected layers or the next stage of the network [33]. To solve this problem, the authors in [34] introduced a new method known as Spatial Pyramid Pooling Network layer (SPP-Net). This layer is designed to overcome the limitations of fixed-size pooling by dividing the feature map into multiple levels of spatial bins with different sizes.

By incorporating the SPP Net layer, R-CNN saw a significant enhancement in speed without compromising the quality of detection. The increase in the speed is because of the reason that the convolutional layer is required to scan only once on the entire image, creating fixed-length characteristics for region proposals of varying sizes.

3.2.3 Fast RCNN

Ross Girshick proposed Fast R-CNN in 2015 for the first time [35]. This is an advanced version of the R-CNN (Region-based Convolutional Neural Network) framework for detection of the objects. It significantly improves both the speed and accuracy of object detection tasks by addressing several limitations of its predecessor. The key challenges with R-CNN and SPP-Net was the necessity to train multiple systems independently. This idea addressed this issue with the development of a single comprehensive trainable model. Instead of using a pyramidal structure of pooling layers (as in Spatial Pyramid Pooling Networks), this model proposed the Region of Interest (RoI) pooling layer. It pools features from the feature maps for each object proposal into fixed-size regions, making it possible to feed these regions into fully connected layers. This layer further extracts features from a specific region of the feature map and normalizes them to a fixed size, allowing the network to manage object proposals of varying sizes. After this RoI pooling layer, the network includes two fully connected layers that process the pooled features, i.e. N + 1-class SoftMax layer and a bounding box regressor layer. In addition, this model replaces the L2 loss function used in R-CNN with Smooth L1 loss for the regression of the bounding box, which reduces the influence of outliers and improves performance.

3.2.4 Faster RCNN

Despite the momentous rise in accuracy and speed achieved by Fast R-CNN, it still relied on the selective search process to generate 2000 region proposals, which was a sluggish process. The authors in [36, 37] addressed this issue by developing a novel detector named Faster R-CNN. This improved the speed of detection by substituting the conventional region proposal algorithms such as selective search [38], or edge boxes [39], multiscale combinatorial grouping [40], with a network called the Region Proposal Network (RPN) [41]. This detector has the four main components explained as below:

(a) The CNN: The output is a feature map which presents the input image in a more abstract form.
(b) The RPN: This is a tiny network that slides over the feature map generated by the backbone and proposes regions (also known as anchors) where objects might be located. These regions are then used for further processing. This generates two outputs i.e. objectness score and bounding box regression.
(c) Region of Interest (RoI) Pooling: The RoIs are pooled to a fixed size so that they can be processed by the next layers regardless of their original size.
(d) Classification and Bounding Box Regression: The pooled RoIs are thereafter allowed to pass across fully connected layers for classification and bounding box refinement.

3.2.5 Mask RCNN

This is an enhancement over Faster R-CNN by incorporating an additional section for pixel-level object instance distribution [42]. This section is a fully connected layer enforced to the RoIs to classify each pixel into segments having minimal computational price. It uses same kind architecture as used in its predecessor for object proposals. The main change is the replacement of the RoIPool layer with the RoIAlign layer to escape the problem of pixel-level misalignment because of spatial quantization. This model uses ResNeXt-101 [43] as the backbone network, along with the Feature Pyramid Network (FPN), to achieve better speed and accuracy. The loss function in this model is also upgraded to mask loss, which employs 5 anchor boxes with 3 aspect ratios, same as in FPN. The training process of both Mask R-CNN and Faster R-CNN is almost same.

This model outperformed the top-tier single-model architectures and is having an additional functionality of instance segmentation with minimal additional computational cost. Further, this model is flexible and simple to train and can be beneficial for purposes like keypoint detection and human pose estimation. Still, it falls short of real-time speed requirement (>30 fps).

4 Data Sets and Performance Metrics

Datasets performs a very significant role when comparing the performance of different type of algorithms. Further, the evaluation of any algorithm can be carried out in terms of certain parameters. Therefore, in this section we will focus on some of the well known datasets and thereafter on the evaluation parameters.

4.1 Data Sets

4.1.1 Pascal VOC

The Pascal Visual Object Classes (VOC) dataset is a prominent and influential dataset in the area of computer vision, particularly for generalized object detection [44]. Developed between 2005 and 2012, the dataset has been extensively employed for evaluating the effectiveness of object detection methods. Pascal VOC 2007 contains twenty Object Classes and 9 k images, whereas Pascal VOC 2012 have approximately 11,000 images divided into twenty classes. VOC 2012 is an enhanced release of VOC 2007 having more images and annotations. This dataset has played a crucial role as a benchmark in the development and evaluation of object detection algorithms. It has been particularly important in the early times of based object detection algorithms based on CNN. The primary metric used for evaluating the performance of any algorithm on the Pascal VOC dataset is Mean Average Precision (mAP) which is explained in next section. Higher mAP values indicate better performance.

The Pascal VOC datasets (2007 and 2012) remain a fundamental resource in the history of computer vision, providing a solid foundation for the research of various object detection techniques.

4.1.2 MS COCO

The Microsoft Common Objects in Context (MS COCO) dataset is a pivotal and arduous massive dataset in the area of computer vision [45]. Developed by Microsoft in 2014, the MS COCO dataset was funded to provide a larger and more diverse set of data for detection and segmentation. This comprises 80 object categories and 330,000 images. Every single image in this dataset is annotated with labels, key information about the object and bounding boxes. The dataset covers a broad range of situations, making it more challenging than many earlier datasets. Many images contain objects that are partially or fully occluded, posing a significant challenge for detection algorithms. A substantial number of small objects are present, requiring algorithms to have high precision and sensitivity. Images often feature dense groups of objects, necessitating advanced algorithms for accurate detection and segmentation. Rich contextual information in the images, reflecting real-world complexity and interactions between objects.

This dataset provides a larger scale and more disparate set of images compared to the Pascal VOC dataset. This dataset is widely used for advancing and testing object detection and segmentation tasks.

4.1.3 ImageNet

It is a significant large-scale standard dataset which has significantly promoted the evolution of the algorithms for object detection [46]. This was initially used for the ImageNet Large Scale Visual Recognition Challenge (ILSVRC) [47] that targets to assess the accomplishments of the study of computer vision on the localizing task and the classification of the objects. This dataset consists of 21,000 classes and each category includes hundreds of thousands of images.

There are other datasets also like OpenImages [48], KITTI, PartNet, CityScapes etc. as mentioned in the survey paper [49], which can be used for the evaluation of the different types of algorithms (Table 2).

Table 2 Datasets summary

Sr. No.	Name of the data set	Year of development	Features
1	Pascal VOC	2005–2012	20,000 images in 20 classes
2	MC COCO	2014	3,30,000 images in 80 object categories
3	ImageNet	2006	1000s images in 21,000 Classes
4	OpenImages	2016–2022	V7 contains 1.9 M images in 600 classes

4.2 Performance Metrics

There are a number of parameters available in the literature that can be applied to evaluate the achievements of any object detection algorithms. Some of the important parameters are explained below.

4.2.1 Accuracy

This is defined as the ratio of the count of the right estimates to the total count of estimates made by the object detection model.

$$Accuracy = \frac{TP + TN}{TP + FP + TN + FN}$$

where the terms

TP (True Positive) = Right estimates of the positive class.

TN (True Negative) = Right estimates of the negative class.

FP (False Positive) = Wrong estimates of the positive class.

FN (False Negative) = Wrong estimates of the negative class.

4.2.2 Precision

This is a parameter utilized to measure the effectiveness of classification algorithms, specifically in situations when the classes are uneven or when the cost of false positives is high. This can be explained as the ratio of true positives to the sum of true positive and false positives.

$$Precision = \frac{TP}{TP + FP}$$

4.2.3 Mean Average Precision (mAP)

It assesses the correctness of a model in recognizing and localizing objects in images. This metric is evaluated by taking the mean of the average precsion values for all classes. In object detection tasks, where models are trained to recognize multiple object classes, mAP provides a single metric to summarize the performance across all classes.

$$mAP = \frac{1}{C}\sum_{i=1}^{C} P_i$$

where P_i = Average precision across each class.

4.2.4 Recall

Recall (also known as sensitivity or true positive rate) is a parameter which measures the model's capability to find all related instances of the positive class. This metric particularly important when missing positive instances can be more costly. It can be explained as the ratio of true positive to the sum of true positive and false negative.

$$Recall = \frac{TP}{TP + FN}$$

4.2.5 F1 Score

This metric is important specifically when dealing with imbalanced datasets. It provides a balance between two metrics i.e. precision and recall, is helpful when both false positive and false negative are significant to consider. It is defined as the harmonic mean of precision and recall.

$$F1\,Score = \frac{2 \times Precision \times Recall}{Precision + Recall}$$

5 Future Research Challenges

Object detection has made significant strides over the past decade. In certain specialized areas, algorithms have nearly achieved human-level accuracy. Despite this progress, there are still many interesting problems to address. In this segment, we will explore some of the interesting research problems in the discipline of object detection.

5.1 Auto ML

The application of automatic neural architecture search (NAS) to determine the features of object detectors is a flourishing field by now. The use of these algorithms have advantages in the detection of small objects and hyper parameter tuning Although some work has been reported but still challenging to find an appropriate algorithm [24, 50, 51].

5.2 Transformers

Although transformers were only recently introduced to computer vision. These have already attained top of the line performance on several benchmarks. The application of transformers especially when combined with CNNs, have demonstrated promising outcomes but demands further exploration [52–54].

5.3 Weakly Supervised Detection

Most of the currently available methods are trained on multitude of bounding box annotated data that is not possible to scale due to the time and resources required for annotation. The capability to train on weakly supervised data, such as image-level labeled data, can possibly reduce these costs in future.

5.4 Scale Adaption

Scale adaptation in object detection relates to techniques that enhance the detection of objects at several scales within an image. This is crucial because objects can appear at different sizes due to variations in distance, perspective, and resolution. The approaches like multi-scale feature fusion, adversial training and Scale-Auxiliary Feature Enhancement could be explored in future for scale adaptation.

5.5 Optimization

The structure of deep convolutional neural networks (DCNNs) can be revamped using various meta-heuristic approaches for betterment. This set of approaches are capable of extending convolutional neural networks in different types of research problems and applications, such as fine-tuning DCNN hyper parameters and DCNN training. Therefore, the appropriateness of meta-heuristic techniques requires further exploration.

5.6 Generative Adversarial Network (GAN) Based Detection

This refers to using Generative Adversarial Networks (GANs) in the discipline of object detection. GANs are a class of machine learning frameworks where two neural networks, a generator and a discriminator, are trained simultaneously through adversarial processes. The generator generates new data samples and discriminator evaluates the authenticity of the samples produced by the generator. These methods are particularly helpful when in real time implementations, images are not very clarified.

In [55], authors have employed several weather augmentation techniques to deal with the images and presented various denoising techniques to improve the performance of state of the art methodologies. Integrating GANs with object detectors has the definite potential of enhancing the reliability of these algorithms in acute circumstances such as partial occlusion, blurring, or other disturbances.

5.7 3D Object Detection

3D object detection becomes a very challenging task specially when applied to autonomous driving. Despite models achieving higher accuracies, deploying anything less than human comparative performance will raise safety issues. This should be the topic of interest for future research specifically for autonomous vehicles.

6 Conclusion

However object detection has made significant advancement in the last 10–15 years, the optimal detectors are yet away from achieving peak results. The demand for low weight models which could be easily implemented for mobile and embedded systems is expected to rise intensively because of its applications in real world. In this paper, we have started the review with the traditional approaches and moved on to the modern approaches. Various types of one stage and two stage detectors have been discussed in terms of their advantages, disadvantages and complexity. The datasets available in the literature and important parameters used for performance monitoring have also been discussed. The future research directions along with the applications has also been discussed, to provide an in-depth coverage of the field. With so much development and positive trend in the area of object detection, still there is scope of improvement specifically for real world applications like autonomous vehicles.

References

1. Peng Y, Qin Y, Tang X, Zhang Z, Deng L (2022) Survey on image and point-cloud fusion-based object detection in autonomous vehicles. IEEE Trans Intell Transp Syst 23(12):22772–22789
2. Wang K, Zhou T, Li X, Ren F (2023) Performance and challenges of 3D object detection methods in complex scenes for autonomous driving. IEEE Trans Intell Veh 8(2):1699–1716
3. Liu P, Wang Z, Yu G, Zhou B, Chen P (2024) Region-based hybrid collaborative perception for connected autonomous vehicles. IEEE Trans Veh Technol 73(3):3119–3128
4. Neto PC, Pinto JR, Boutros F, Damer N, Sequeira AF, Cardoso JS (2022) Beyond masks: on the generalization of masked face recognition models to occluded face recognition. IEEE Access 10:86222–86233
5. Huang Z, Zhang J, Shan H (2023) When age-invariant face recognition meets face age synthesis: a multi-task learning framework and a new benchmark. IEEE Trans Pattern Anal Mach Intell 45(6):7917–7932
6. Li N et al (2023) Chinese face dataset for face recognition in an uncontrolled classroom environment. IEEE Access 11:86963–86976
7. Park S, Na H, Choi D (2024) Verifiable facial de-identification in video surveillance. IEEE Access 12:67758–67771
8. Dılek E, Dener M (2024) Enhancement of video anomaly detection performance using transfer learning and fine-tuning. IEEE Access 12:73304–73322
9. Safwat S, Mahmoud A, Eldesouky Fattoh I, Ali F (2024) Hybrid deep learning model based on GAN and RESNET for detecting fake faces. IEEE Access 12:86391–86402
10. Jiang P, Xie H, Yu L, Jin G, Zhang Y (2024) Exploring bi-level inconsistency via blended images for generalizable face forgery detection. IEEE Trans Inf Forensics Secur 19:6573–6588
11. Jiao L et al (2022) New generation deep learning for video object detection: a survey. IEEE Trans Neural Netw Learn Syst 33(8):3195–3215
12. Rajawat D, Lohani BP, Rana A, Srivastava A, Yadav P, Gupta S (2023) Object detection in images and videos using OpenCV: a comparative study of deep learning and traditional computer vision techniques. In: 10th IEEE Uttar Pradesh section international conference on electrical, electronics and computer engineering (UPCON), Gautam Buddha Nagar, India, pp 141–146
13. Ye T, Qin W, Zhao Z, Gao X, Deng X, Ouyang Y (2023) Real-time object detection network in UAV-vision based on CNN and transformer. IEEE Trans Instrum Meas 72:1–13
14. Ibrahim E, Zaghden N, Mejdoub M (2024) Semantic analysis system to recognize moving objects by using a deep learning model. IEEE Access 12:80740–80753
15. Viola P, Jones M (2001) Rapid object detection using a boosted cascade of simple features. In: Proceedings of the 2001 IEEE computer society conference on computer vision and pattern recognition. CVPR 2001, Kauai, HI, USA, pp I–I
16. Dalal N, Triggs B (2005) Histograms of oriented gradients for human detection. In: IEEE computer society conference on computer vision and pattern recognition (CVPR'05), San Diego, CA, USA, vol 1, pp 886–893
17. Felzenszwalb P, McAllester D, Ramanan D (2008) A discriminatively trained, multiscale, deformable part model. In: IEEE conference on computer vision and pattern recognition, Anchorage, AK, USA, pp 1–8
18. Jiao L et al (2019) A survey of deep learning-based object detection. IEEE Access 7:128837–128868
19. Sermanet P et al (2013) OverFeat: integrated recognition, localization and detection using convolutional networks. arXiv:1312.6229
20. Lin T-Y, Dollár P, Girshick R, He K, Hariharan B, Belongie S (2017) Feature pyramid networks for object detection. In: IEEE conference on computer vision and pattern recognition (CVPR), Honolulu, HI, USA, pp 936–944
21. Lin T-Y, Goyal P, Girshick R, He K, Dollár P (2017) Focal loss for dense object detection. In: IEEE international conference on computer vision (ICCV), Venice, Italy, pp 2999–3007

22. Liu et al (2016) SSD: single shot multibox detector. In: 14th European conference on computer vision, ECCV, Amsterdam
23. Redmon J, Divvala S, Girshick R, Farhadi A (2016) You only look once: unified, real-time object detection. In: Proceedings of the IEEE conference on computer vision and pattern recognition, pp 779–788
24. Liu L, Ouyang W, Wang X, Fieguth P, Chen J, Liu X, Pietikäinen M (2020) Deep learning for generic object detection: a survey. Int J Comput Vis 128:261–318
25. Redmon J, Farhadi A (2017) Yolo9000: better, faster, stronger. In: Proceedings of the IEEE conference on computer vision and pattern recognition, pp 7263–7271
26. Redmon J, Farhadi A (2018) Yolov3: an incremental improvement. arXiv:1804.02767
27. Bochkovskiy A, Wang C-Y, Liao H-YM (2020) YOLOv4: optimal speed and accuracy of object detection, April 2020. arXiv:2004.10934
28. Jocher G et al (2021) ultralytics/yolov5: v6.0 -YOLOv5n 'Nano' models, Roboflow integration, TensorFlow export, Open CV DNN support, October 2021. https://doi.org/10.5281/zenodo.5563715
29. Solawetz J (2022) YOLOv5 new version -improvements and evaluation, June 2020. https://blog.roboflow.com/yolov5-improvements-and-evaluation/. Accessed 1 Apr 2022
30. Girshick R, Donahue J, Darrell T, Malik J (2014) Rich feature hierarchies for accurate object detection and semantic segmentation. In: Proceedings of the IEEE conference on computer vision and pattern recognition (CVPR)
31. Uijlings JRR, van de Sande K, Smeulders Gevers T, Smeulders AWM (2013) UvA-DARE (digital academic repository) selective search for object recognition selective search for object recognition
32. Krizhevsky A, Sutskever I, Hinton GE (2012) ImageNet classification with deep convolutional neural networks. In: Pereira F, Burges CJC, Bottou L, Weinberger KQ (eds) Advances in neural information processing systems. Curran Associates, Inc., p 9
33. Park S (2021) A guide to two-stage object detection: R-CNN, FPN, mask R-CNN, July 2021. https://medium.com/codex/a-guide-to-two-stage-object-detection-r-cnn-fpn-mask-r-cnn-and-more-54c2e168438c
34. He K, Zhang X, Ren S, Sun J (2015) Spatial pyramid pooling in deep convolutional networks for visual recognition. IEEE Trans Pattern Anal Mach Intell 37(9):1904–1916
35. Girshick R (2015) Fast R-CNN. In: IEEE international conference on computer vision (ICCV), pp 1440–1448
36. Ren S, He K, Girshick R, Sun J (2015) Faster r-cnn: towards real-time object detection with region proposal networks. Adv Neural Inf Process Syst 28
37. Ren S, He K, Girshick R, Sun J (2017) Faster R-CNN: towards real-time object detection with region proposal networks. IEEE Trans Pattern Anal Mach Intell 39(6):1137–1149
38. Uijlings JR, Van De Sande KE, Gevers T, Smeulders AW (2013) Selective search for object recognition. Int J Comput Vis 104(2):154–171
39. Zitnick CL, Dollár P (2014) Edge boxes: locating object proposals from edges. In: European conference on computer vision. Springer, pp 391–405
40. Arbeláez P, Pont-Tuset J, Barron JT, Marques F, Malik J (2014) Multiscale combinatorial grouping. In: Proceedings of the IEEE conference on computer vision and pattern recognition, pp 328–335
41. Kaur R, Singh S (2023) A comprehensive review of object detection with deep learning. Digit Signal Process Springer 132:103812
42. He K, Gkioxari G, Dollár P, Girshick R (2018) Mask R-CNN. arXiv:1703.06870
43. Xie S, Girshick R, Dollár P, Tu Z, He K (2016) Aggregated residual transformations for deep neural networks. arXiv:1611.05431
44. Everingham M, Van Gool L, Williams CK, Winn J, Zisserman A (2010) The pascal visual object classes (VOC) challenge. Int J Comput Vis 88(2):303–338
45. T.-Y. Lin, M. Maire, S. Belongie, J. Hays, P. Perona, D. Ramanan, P. Dollár, C.L. Zitnick, Microsoft coco: common objects in context, in: European Conference on Computer Vision, Springer, pp.740–755, 2014.

46. J. Deng, W. Dong, R. Socher, L.-J. Li, K. Li, L. "Fei-Fei, Imagenet: a large-scale hierarchical image database." IEEE Conference on Computer Vision and Pattern Recognition, IEEE, pp.248–255, 2009.
47. Russakovsky O, Deng J, Su H, Krause J, Satheesh S, Ma S, Huang Z, Karpathy A, Khosla A, Bernstein M et al (2015) Imagenet large scale visual recog-nition challenge. Int J Comput Vis 115(3):211–252
48. Kuznetsova A, Rom H, Alldrin N, Uijlings J, Krasin I, Pont-Tuset J, Kamali S, Popov S, Malloci M, Kolesnikov A et al (2020) The open images dataset v4. Int J Comput Vis 128(7):1956–1981
49. Sun Y, Sun Z, Chen W (2024) The evolution of object detection methods. Eng Appl Artif Intell 133(Part E):108458
50. Heller M (2022) What is neural architecture search? AutoML for deep learning, January 2022. https://www.infoworld.com/article/3648408/what-is-neural-architecture-search.html. Accessed 26 Feb 2022
51. Everything you need to know about AutoML and neural architecture search, https://www.kdnuggets.com/2018/09/everything-need-know-about-automl-neural-architecture-search.html. Accessed 26 Feb 2022
52. Xiao T, Singh M, Mintun E, Darrell T, Dollár P, Girshick R (2021) Early convolutions help transformers see better. arXiv:2106.14881
53. Wu H, Xiao B, Codella N, Liu M, Dai X, Yuan L, Zhang L (2021) CvT: introducing convolutions to vision transformers. arXiv:2103.15808
54. d'Ascoli S, Touvron H, Leavitt M, Morcos A, Biroli G, Sagun L (2021) ConViT: im-proving vision transformers with soft convolutional inductive biases. arXiv:2103.10697
55. Gupta H, Kotlyar O, Andreasson H, Lilienthal AJ (2024) Robust object detection in challenging weather conditions. In: Proceedings of the IEEE/CVF winter conference on applications of computer vision, pp 7523–7532

Unlocking the Secrets of Deep Space: AI Techniques for Object Detection in Astronomical Imagery

Manjuleshwar Panda and Yogesh Chandra

Abstract The exploration of deep space has always been a source of fascination for humans, with astronomical photography providing crucial insights into the enormous expanse of the cosmos. The study explores artificial intelligence's transformational role in furthering astronomical science. It begins by emphasizing the importance and problems of recognising celestial objects in complicated and massive astronomical data. The basic ideas of machine learning and artificial intelligence (AI) are then briefly discussed, emphasizing how they are used in object detection and picture processing. The key preprocessing approaches for improving the quality and usability of astronomical photographs are thoroughly explored. The work examines different AI approaches, including Convolutional Neural Networks (CNNs), Transfer Learning, and advanced models like YOLO and Mask R-CNN, to demonstrate their usefulness in finding and categorizing celestial events. Real-world case studies demonstrate how these techniques can be used to detect exoplanets, galaxies, and supernovae and contribute to gravitational wave research and radio astronomy. Looking ahead, the chapter discusses potential directions, including advances in AI algorithms, integration with robotic telescopes, ethical implications, and obstacles like data constraints and bias. This chapter hopes to inspire further advances in the quest to understand the universe by bridging the gap between cutting-edge AI technology and astronomical exploration.

Keywords Astronomical imagery · Object detection · Artificial intelligence · Deep learning · Convolutional Neural Networks (CNNs)

Humanity has always been captivated by the cosmos, which calls us to solve its enigmas and appreciate its immense complexity. Our window into the cosmos is provided by astronomical photography, which allows us to see far-off galaxies,

M. Panda
Delhi, India

Y. Chandra (✉)
Department of Physics, Government P.G. College Bazpur, Bazpur, Uttarakhand, India
e-mail: yepphysics@gmail.com

S. Chand Kumain et al. (eds.), *AI and ML Techniques in Image Processing and Object Detection*, https://doi.org/10.1007/978-981-96-7445-9_4

decipher celestial occurrences, and investigate the complex fabric of deep space. However, processing and analyzing these enormous datasets has become more difficult than ever due to the exponential expansion in astronomical data brought on by advancements in telescopes and space missions. Artificial intelligence (AI) is a disruptive force that is revolutionizing the way we discover, classify, and interpret astronomical objects in this data-rich era. The nexus between artificial intelligence and astronomy illuminates the state-of-the-art methods and tools that are revolutionizing how we approach the investigation of the universe's most remote regions. With the combination of astronomy and artificial intelligence, the impossible is becoming inevitable in a cosmos full of undiscovered marvels.

1 Introduction to Astronomical Object Detection

Astronomical object detection is the process of recognising and cataloging celestial phenomena such as stars, galaxies, and asteroids using astronomical imagery. This process is required for comprehension of the universe's structure and dynamics. Modern telescopes produce massive volumes of data, creating both opportunities and challenges. The intricacy and scale of this data necessitate the use of advanced analysis techniques. Artificial intelligence (AI) and machine learning (ML) have become indispensable tools in this domain, revolutionizing our analysis of cosmic pictures and opening up new avenues for research.

1.1 Importance of Object Detection in Astronomy

Astronomical object identification is critical to expanding our understanding of the universe. Astronomers can learn about the genesis, evolution, and interaction of celestial bodies including stars, galaxies, nebulae, and exoplanets by detecting and cataloging them. This procedure is required for producing detailed sky maps and surveys, which aid in the study of cosmic phenomena and comprehending the structure of the cosmos. Furthermore, object detection makes it possible to identify brief and unusual occurrences like gravitational waves, supernovae, and near-Earth asteroids—all of which have important scientific and practical ramifications [1]. Detecting asteroids, for example, can help assess possible Earth-impact hazards, whereas spotting supernovae helps us comprehend stellar life cycles and the expansion of the universe.

1.2 Challenges of Analyzing Astronomical Imagery

Analyzing astronomical images is difficult due to the large data volumes and the necessity for high precision. Gigabytes of data are produced by modern telescopes like the Hubble Space Telescope (HST) and the James Webb Space Telescope (JWST), making manual analysis impractical. Furthermore, images are frequently corrupted with noise and artifacts from a variety of causes, including atmospheric distortion, instrumentation mistakes, and cosmic ray impacts. These conditions can hide or imitate celestial objects, making them difficult to detect and classify. To ensure accuracy and dependability, differentiate actual astronomical objects from false positives using complex algorithms and robust validation procedures [2].

1.3 An Overview of AI and Machine Learning in Astronomy

Astronomical object recognition has changed as a result of artificial intelligence (AI) and machine learning (ML), which have produced strong tools for handling big datasets and complex patterns. The identification and classification of celestial objects can be automated with the use of artificial intelligence and machine learning methods, greatly accelerating research. An example of a deep learning model that is particularly well-suited for astronomical image analysis is convolutional neural networks (CNNs), which are excellent at image recognition tasks. These models have been successfully used to classify galaxies, find exoplanets, and identify gravitational lenses [3]. In addition to improving object recognition's effectiveness and accuracy, AI and machine learning can open up new avenues for research by spotting connections and patterns that traditional techniques would have missed.

In conjunction with these advancements, the issue of categorization accuracy has become central to AI-powered astronomical analysis. The ratio of successfully categorized objects to the total number of objects analyzed determines the classification accuracy and can be stated in Eq. (1), as follows:

$$\begin{aligned} Classification Accuracy &= (Quantity\ of\ Accurate\ \mathrm{Pr}\,edictions)/ \\ &\quad (Total\ Number\ of\ Forecasts) \end{aligned} \tag{1}$$

This simple formula gives a clear indication of how well AI and ML systems perform in tasks like recognising and classifying celestial objects. For example, if an AI model examines 1,000 images of galaxies and properly identifies 950 of them, the classification accuracy is 95%. This simple metric assists astronomers in determining the dependability of AI models in a variety of applications, including finding exoplanets and discriminating between different types of galaxies. By boosting classification accuracy using techniques such as CNNs and combining massive datasets, AI not only improves the efficiency of astronomical research, but also enables new discoveries by identifying patterns and connections that traditional methods may miss.

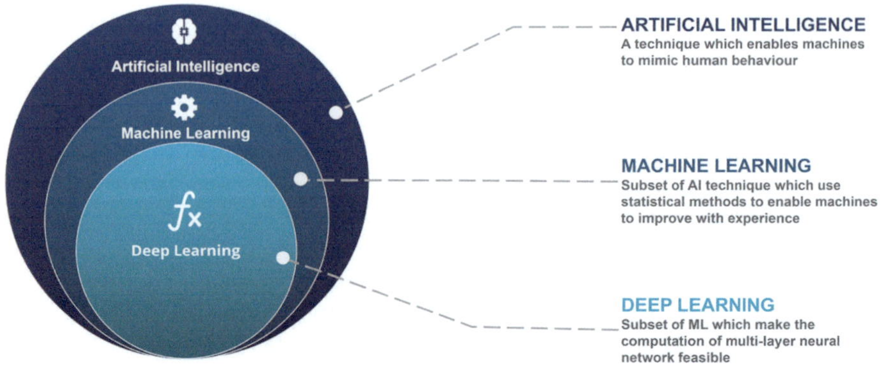

Fig. 1 DL < ML < AI (*Credit* Edureka)

2 Fundamentals of AI and Machine Learning

2.1 Basic Concepts of AI and Machine Learning

AI and ML are cutting-edge techniques that let computers mimic human intellect and learn from information. Artificial Intelligence (AI) comprises a wide range of methods, such as robotics, rule-based systems, and natural language processing. Within the field of artificial intelligence, machine learning focuses on creating algorithms that allow computers to recognise patterns in data and make choices or predictions without needing to be explicitly programmed for each task.

These algorithms (See Fig. 1) can handle and analyze massive volumes of data quickly and efficiently, making them useful in fields that produce enormous datasets, such as astronomy [4]. AI and machine learning have proven especially useful in astronomy, where the capacity to handle large volumes of data quickly can lead to ground-breaking discoveries like detecting previously undiscovered celestial objects. These technologies are not only improving our understanding of the universe, but they are also opening up new research opportunities by revealing patterns and correlations that were previously unknown to us.

2.2 Supervised Versus Unsupervised Learning

The two primary categories of machine learning are supervised and unsupervised learning. Supervised learning involves training algorithms on labeled datasets in which the input data matches the proper output. For problems like regression and classification, this method is employed. For example, in the field of astronomical object detection, a set of labeled images, each with an annotation identifying the kind of celestial object it includes, could be used to train a supervised learning model.

The model learns to recognise the traits associated with each object category and can then categorize new, unlabelled photos [5]. Some key points about supervised learning are as follows:

- **Labeled Data Requirement**: Supervised learning is based on huge datasets, with each input coupled with a known outcome, providing a clear reference for the system to learn from.
- **Prediction Accuracy**: When the model is trained on large datasets, it produces predictions that are incredibly accurate because its efficacy is directly correlated with the caliber and volume of labeled data.
- **Error Adjustment**: The model's parameters are adjusted during training by comparing the model's predictions with the actual labels; this allows the model to perform better over time.
- **Broad Applicability**: Supervised learning is widely applied in many fields, such as astronomy, where it is used to classify objects in the sky, forecast the rates at which stars originate, and identify different types of galaxies using data that has already been labeled.

The training of algorithms on data without labeled outputs is known as unsupervised learning, in contrast. Finding hidden structures or patterns in the data is the aim. In unsupervised learning, dimensionality reduction and grouping are two popular strategies. To combine galaxies with similar forms or spectral signatures, for example, or to classify similar objects based on their properties, astronomers may use unsupervised learning. This method can reveal previously unknown insights and linkages in the data. Key Points include as follows:

- **Lack of Labeled Data**: Unsupervised learning operates without the need for labeled outputs, enabling the algorithm to investigate and identify patterns just in the input data.
- **Pattern Recognition**: This method is especially good at finding natural groups or hidden structures in the data, such as grouping stars according to their inherent characteristics.
- **Flexibility**: Unsupervised learning is very helpful for analyzing complex or unknown astronomical datasets because of its great versatility and ability to be applied to a wide range of data sources.
- **Insight Generation**: Unsupervised learning can provide new astronomical findings and theories by revealing links and patterns that were previously unknown, deepening our grasp of the cosmos.

2.3 Deep Learning and Neural Network

Deep learning, as a branch of machine learning, gets its name from the fact that it models complicated patterns in data by using neural networks with numerous layers (hence the name "deep"). Neural networks are modeled after the human brain's structure, which consists of interconnected nodes (neurons) that process information.

Table 1 Comparison sheet for deep learning versus neural networks

Parameters of comparison	Deep learning	Neural networks
Layers	Multiple	Single or Few
Complexity	High	Moderate
Learning compatibilities	Advanced feature extraction and abstraction	Learning of fundamental features
Training data	Needs substantial datasets	Able to operate with smaller datasets
Use cases	NLP, astronomical detection, and image recognition	Simple challenges involving regression and categorization

The model can comprehend intricate relationships and carry out tasks like image recognition and natural language processing with a high degree of accuracy because each layer of the network pulls increasingly more abstract elements from the input data [6].

Deep learning models that function particularly well for image processing are called convolutional neural networks (CNNs). Convolutional layers are used by CNNs to automatically deduce the hierarchies of spatial features from images. This makes them excellent for astronomical object detection, as they can be trained to recognize specific celestial objects based on attributes like shape, brightness, and texture. Deep learning models have improved our understanding of the universe and made it easier to uncover new celestial events by boosting the correctness and ability of object detection in astronomical imagery. Here is the comparison in Table 1, as follows:

Astronomy has undergone a revolution because of the incorporation of deep learning, especially with models like CNNs, which enable more precise and effective interpretation of celestial imagery. A deeper understanding of the universe and fresh discoveries are being made possible by this technological breakthrough.

3 Preprocessing Astronomical Images

3.1 Data Acquisition and Sources for Astronomical Images

The core of any successful astronomical object detection project is high-quality data collection. Very Large Telescopes (VLT) in Chile and space-based observatories like the Hubble Space Telescope (HST) and the James Webb Space Telescope are two popular tools used to take astronomical images. These telescopes collect massive volumes of data at multiple wavelengths, ranging from visible light to infrared and X-rays. Enormous datasets for research and the growth of sophisticated AI and machine learning algorithms are provided via publicly available databases, incl. the NASA Exoplanet Archive and the Sloan Digital Sky Survey (SDSS) [7].

3.2 Noise Reduction Techniques

Noise in astronomical photographs can come from a variety of sources, including the Earth's atmosphere, thermal fluctuations in detectors, and cosmic radiation. Effective noise reduction is required to improve image quality and enable proper object detection. Noise reduction techniques include median filtering, Gaussian smoothing, and wavelet transforms. Median filtering effectively eliminates impulsive noise, but Gaussian smoothing minimizes Gaussian noise by averaging pixel values with a Gaussian kernel [8]. The wavelet transform is very useful in denoising astronomical photographs because it divides the image into different frequency components, allowing for the elimination of noise while keeping significant characteristics. There are several types of image noise filters. They are normally classified into two categories: time domain and frequency domain. Their brief description is following:

Time Domain Filters:

1. **Median filter**:
   - A non-linear filter that uses the median of the pixels next to each one to replace the value of each pixel.
   - Highly helpful at eliminating impulsive or salt-and-pepper' noises.
   - Edges are better preserved than when using linear filters.
2. **Mean filter**:
   - A linear filter that replaces each pixel's value with the average of the adjacent pixel values.
   - Effective in decreasing random noise, but may blur edges and fine details.
3. **Adaptive filters**:
   - These filters modify their behavior in response to local image properties.
   - Can efficiently minimize noise while maintaining edges and fine details.

Frequency Domain Filters:

1. **Gaussian Filters**:
   - Applied in the frequency domain with a Gaussian kernel.
   - Reduces high-frequency noise while keeping the overall image smooth.
   - Frequently used for Gaussian noise reduction.
2. **Wiener Filter**:
   - The goal of a linear filter is to lower the mean squared deviation between the initial image and the predicted image.
   - Effectively reduces additive noise while keeping crucial image elements.
3. **Wavelet Transform**:
   - Divides the image into many frequency components.

- Allows for customized noise reduction by adjusting certain frequency bands.
- Effective in denoising while retaining important image features, particularly with astronomical images.

These various filtering approaches, both in time and frequency domains, contribute to the improvement of the quality of astronomical images by effectively decreasing noise while maintaining vital information, allowing for more precise object detection and analysis.

3.3 *Image Calibration and Alignment*

Accurate picture calibration and alignment are crucial preparation processes for astronomical photos. Calibration entails adjusting images for instrumental effects such as bias, dark current, and flat fielding. Bias correction eliminates electronic noise introduced by the detector, dark current correction tackles thermal noise, and flat-fielding corrects fluctuations in pixel sensitivity. Once calibrated, pictures must be aligned to ensure that celestial objects are correctly superimposed across different exposures or measurements. Images are aligned with sub-pixel accuracy using techniques like cross-correlation and feature matching. Proper calibration and alignment are required for high-fidelity composite images and subsequent analysis, such as object detection and photometry [9]. This process can be mathematically formalized to enhance the grasp of the procedures engaged in converting raw astronomical data into high-quality images for study. Here's a single, simple formalism in Eq. (2), that summarizes the entire process:

$$I_{final}(x', y') = \left[(I_{raw}(x, y) - B(x, y) - D(x, y))/F(x, y)\right] \times \left(T\left(x', y'\right)\right) \quad (2)$$

Here:

- $\mathbf{I_{raw}(x,y)}$ is the raw image data.
- $\mathbf{B(x, y)}$ is the bias frame, which accounts for electrical noise.
- $\mathbf{D(x, y)}$ represents the dark current frame, which removes thermal noise.
- $\mathbf{F(x,y)}$ is the flat-field frame, which corrects for pixel sensitivity differences.
- $\mathbf{T(x', y')}$ is the transformation matrix used to align the image with sub-pixel accuracy.

This simple formalism combines all of the major phases of image calibration and alignment into a single equation, streamlining the process and making it easier to understand. It emphasizes the sequential application of each correction, resulting in a final, high-quality image suitable for accurate scientific analysis. This method provides a clear mathematical depiction of the calibration and alignment process, highlighting its importance in astronomical imaging.

4 AI Techniques for Object Detection in Astronomy

4.1 Convolutional Neural Networks (CNNs) for Object Detection

Image processing has been transformed by Convolutional Neural Networks (CNNs) and they have demonstrated remarkable success in astronomical object detection. CNNs learn hierarchical features from raw images using many layers of convolutions, pooling, and nonlinear activations. This enables them to detect and classify celestial objects like stars, galaxies, and supernovae with great precision. The Hubble Space Telescope (HST) and the Sloan Digital Sky Survey (SDSS) can effectively gather data with the use of CNNs, improving our ability to analyze large datasets [7]. CNNs, for example, have been used to locate gravitational lenses, which are important in the research of dark matter and universe expansion.

As portrayed in the diagram (refer to Fig. 2), a CNN design is composed of both feature extraction and classification. Multiple convolutional and pooling layers are used to extract features from the input image. The convolutional layers are responsible for identifying local patterns like edges or textures, by applying filters throughout the input image. The spatial dimensions of the data are then reduced by pooling layers, which successfully summarizes the existence of features found by the convolutional layers. This aids in the management of computational resources and lowers the likelihood of overfitting. The fully linked layers are used to send the features to the classification phase after extraction. In this instance, the network integrates the extracted features to create a high-level comprehension of the picture, finally generating an output that symbolizes the categorization of the input image—for example, determining whether a star or a galaxy is there. CNNs are especially effective for astronomical object recognition because of their ability to combine feature extraction with classification, which allows for the highly accurate identification of intricate celestial patterns and structures. The procedure is essential for evaluating enormous volumes of data from observatories and has proven helpful in the advancement of astronomy research.

4.2 Transfer Learning in Astronomy

Transfer learning is the process of fine-tuning a pre-trained model, which is typically learned on huge datasets such as ImageNet, for specific astronomical objectives. This method is very beneficial in astronomy, where labeled data may be limited. Transfer learning, which draws on information from other fields, can greatly increase model performance on astronomical datasets. For example, Fine-tuning is possible for pre-trained CNNs to recognise certain features in Hubble Space Telescope or other observatory photos, preventing the necessity for a significant amount of training data and processing resources. The transfer learning approach can be represented in

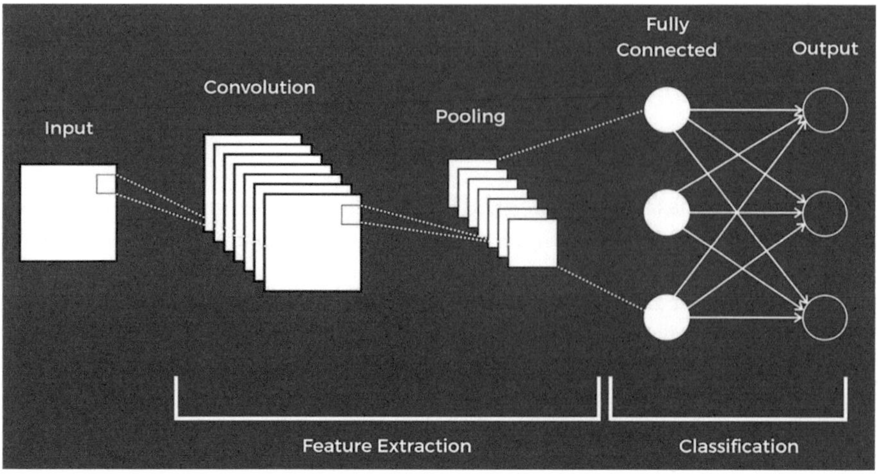

Fig. 2 CNN framework for image classification (*Credit* The Click Reader Team. "Introduction to Convolutional Neural Networks." https://www.theclickreader.com/introduction-to-convolutional-neural-networks/)

Eq. (3), as follows in order to quantify this process:

$$Model_{fine-tuned} = Train\left(Model_{pre-trained}, Data_{specific}, Objective_{task}\right) \quad (3)$$

where:

- $\mathbf{Model_{pre\text{-}trained}}$ represents the original model that has learnt to extract general features from images after being trained on a sizable, all-purpose dataset (such as ImageNet).
- $\mathbf{Data_{specific}}$ is the particular astronomical dataset that is utilized to refine the pre-trained model, such as photos from the Hubble Space Telescope. It includes instances with labels that are pertinent to the astronomical objectives.
- $\mathbf{Objective_{task}}$ indicates the specific objective (such as feature detection or classification) for which the model is being modified in the field of astronomy.
- **Train** (·) function is the procedure that updates the pre-trained model's weights based on the particular data to maximize performance for the assigned task.

Applying this algorithm improves the pre-trained model's conduct on specialized astronomical functions with restricted labeled data by adapting it through training on domain-specific data.

4.3 Region-Based Convolutional Neural Networks (R-CNN)

Region-Based Convolutional Neural Networks (R-CNN) enhance the capabilities of traditional CNNs by including region recommendations for object detection. This method enables the identification of items at different scales and places within an image. R-CNNs have been used to detect and categorize many objects in complex astronomical sceneries, such as densely packed star fields or galaxy clusters. R-CNNs may accurately localize and classify items despite considerable background noise and overlapping objects by producing area suggestions and applying CNNs to them.

4.4 YOLO (You Only Look Once) and Its Variants

YOLO (You Only Look Once) is an object recognition system that identifies images by dividing them into grid cells and predicting bounding boxes and class probabilities with a single forward pass. Real-time object detection applications can benefit greatly from this technique due to its speed. YOLO and its variations, such as YOLOv3 and YOLOv4, have been applied to numerous astronomical datasets, allowing for the quick detection of transitory phenomena like supernovae and gamma-ray bursts. YOLO's speed and precision make it ideal for analyzing massive amounts of astronomical data in real time [10]. In terms of astronomical object detection, YOLO's salient features include:

- **Processing in real-time**: Because YOLO can process photos in real-time, it is very useful for real-time detection of transitory astronomical events, such as rapid radio bursts, gamma-ray bursts, and supernovae.
- **One Forward Pass Only**: Large-scale sky surveys can benefit from YOLO's architecture, which reduces processing time by allowing object detection in a single forward run over the network.
- **Elevated Accuracy**: Even with its rapid speed, YOLO is able to identify and localize objects with great accuracy, guaranteeing the reliable detection of small or faint astronomical occurrences.
- **Flexibility Throughout Datasets**: YOLO and its variations have shown effective when applied to various astronomical datasets, this includes observation of radio and X-rays, along with optical and infrared images.
- **Scalability**: YOLO is scalable for use in large-scale astronomical surveys, where millions of photos may need to be analyzed because of its efficiency in handling enormous volumes of data.
- **Flexibility**: Because of its architecture, YOLO may be optimized for certain astronomical tasks, resulting in optimal performance for a range of observational campaigns and data types.

Since it can quickly and accurately detect celestial occurrences across a variety of datasets, YOLO is a highly valuable tool in modern astronomy.

4.5 MMask R-CNN for Detection and Segmentation

The R-CNN framework is enhanced by Mask R-CNN, which includes a branch that forecasts segmentation masks for every identified object. This is capable of detecting objects and segmenting instances, which are required for full analysis of complicated astronomical images. Mask R-CNN has been used to segment and categorize objects in dense star clusters, as well as to extract individual galaxies from crowded fields. Mask R-CNN provides pixel-level segmentation, allowing for more precise measurements of object attributes including as form, size, and brightness, which improves our understanding of underlying astrophysical processes [11]. Mask R-CNN covers two primary categories of image segmentation:

1. **Semantic Segmentation**:

   - **Description**: The process of classifying each pixel in an image into a preset category while maintaining consistency across different instances of the same category is known as semantic segmentation. The middle image of the given image in Fig. 3 displays the semantic segmentation result, which indicates that the table and all of the chairs have different colors allocated to them according to their categories (e.g., all of the chairs are blue and the table is white). Nevertheless, the model treats every chair as a single entity and does not distinguish between distinct chairs.
   - **Astronomical Application**: Astronomers can examine the overall composition of the image by using semantic segmentation to distinguish between distinct celestial features like stars, galaxies, and nebulae in astronomical images.

2. **Instance Segmentation**:

   - **Description**: By identifying every pixel and differentiating between various instances of the same object class, instance segmentation goes one step further. As can be seen in the instance segmentation output (right image), every chair

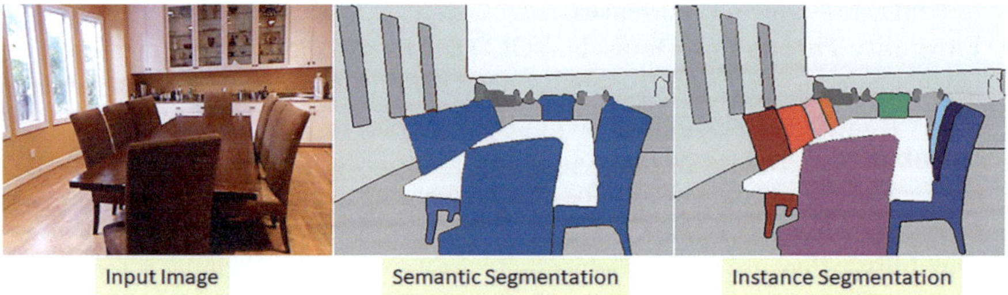

Fig. 3 Disparities between instance and semantic segmentation (*Credit* Medium/TDS. "Review: DeepMask Instance Segmentation." https://towardsdatascience.com/review-deepmask-instance-segmentation-30327a072339)

has a distinct color given to it, making it possible to identify and examine individual objects within the same category.
- **Astronomical Application**: In astronomy, instance segmentation is essential for tasks like separating many stars in a cluttered field or recognizing individual galaxies inside a dense cluster, enabling a more thorough examination of each object.

Mask R-CNN is a very useful tool in astronomy since it can conduct both semantic and instance segmentation. Mask R-CNN helps in the identification and classification of astronomical phenomena and allows for accurate measurement and analysis of celestial objects by offering pixel-level segmentation. This enhances our understanding of intricate astrophysical processes.

5 Case Studies and Applications

5.1 Exoplanet Detection Using Artificial Intelligence

Artificial intelligence has substantially improved the discovery of exoplanets, or planets outside our solar system. Traditional approaches, like the transit method, rely on detecting a star's minute dimming when a planet passes in front of it.

AI algorithms, particularly deep learning models, can analyze light curves from observatories such as Kepler and TESS (See Fig. 4) with surprising accuracy. These models are trained to recognise the tiny signs of exoplanets in the middle of stellar noise, resulting in the finding of countless new planets that would have gone undetected by traditional approaches [12]. For example, the utilization of neural networks has resulted in the identification of exoplanet candidates in data sets including millions of stars, speeding the rate of discovery and expanding our understanding of planetary systems. Building on these developments, artificial intelligence (AI) has revolutionized the search for exoplanets, especially when applied to data from important space missions.

- **Kepler Mission**: By continually observing the brightness of more than 150,000 stars and looking for the telltale dimming that happens when a planet comes between itself and its host star, the Kepler Space Telescope transformed the search for exoplanets. Kepler's enormous data sets have proven to be a valuable resource for sorting through using AI techniques, especially deep learning models. These algorithms are taught to identify, even in the presence of strong stellar noise, the tiny patterns linked to planetary transits. AI has automated the detection process and improved the accuracy of results, leading to the discovery of numerous extrasolar planets, some of which are Earth-sized in the habitable zone.
- **TESS Mission**: Kepler's legacy is furthered by the Transiting Exoplanet Survey Satellite (TESS), which concentrates on the brightest stars while surveying almost

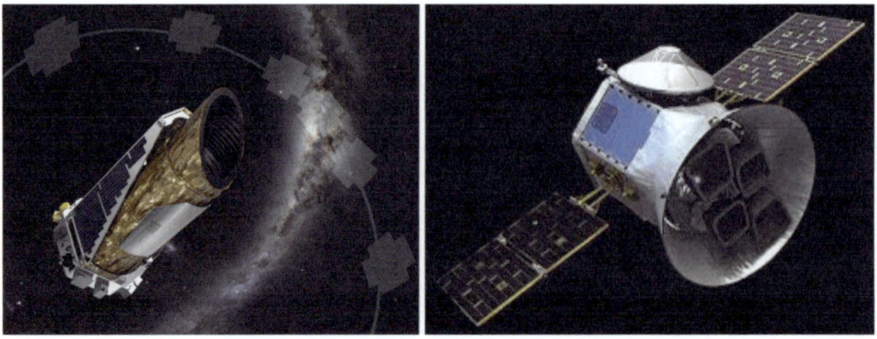

Fig. 4 Kepler space telescope—Kepler (left) and transiting exoplanet survey satellite—TESS (right) (*Credit* NASA)

the whole sky. TESS generates a significant amount of data, which AI models effectively process to find new exoplanets. Smaller and farther-off exoplanets that could have gone unnoticed by traditional approaches can now be found thanks to AI-driven algorithms that are able to detect the minute dips in sunlight created by transiting planets. A key factor in quickening the rate of exoplanet discoveries and expanding our understanding of planetary diversity is the use of AI in TESS data processing.

Our knowledge of the variety and distribution of planetary systems in the galaxy has grown as a result of the incorporation of AI into the analysis of data from Kepler and TESS, both of which have improved the speed and efficiency of exoplanet discovery. With artificial intelligence playing a major part in the quest for the hidden planets that form our universe, these developments usher in a new age in exoplanet research.

5.2 *Identifying Galaxies and Star Clusters*

AI approaches have transformed the identification and classification of galaxies and star clusters. Convolutional Neural Networks (CNNs) are particularly excellent at processing large volumes of celestial imagery, distinguishing between different types of galaxies, such as spiral, elliptical, and irregular, as well as identifying star clusters. Researchers can automate the classification process by training these models on labeled information from the Sloan Digital Sky Survey (SDSS) is just among many surveys, which reduces manual labor while enhancing accuracy. Furthermore, AI models have helped uncover new star clusters in crowded regions when traditional methods fail to discern individual objects due to overlapping light sources [13].

5.3 *Supernova Detection and Classification*

Supernovae, or stellar explosions, are crucial for comprehending the universe's dynamics. AI has significantly improved supernova detection and categorisation by analyzing data from automated sky surveys. By analyzing vast volumes of data, machine learning algorithms can uncover and classify transitory events using light curves and spectra. Random Forests and Support Vector Machines (SVMs) have been used to discriminate between distinct types of supernovae, such as Type Ia and Type II, thereby advancing the study of stellar evolution and cosmology. AI-powered technologies enable the rapid identification of supernova candidates, making follow-up observations necessary for further investigation. Although they have different advantages, Random Forests and Support Vector Machines (SVMs) are both efficient machine learning algorithms for classifying different kinds of supernovae.

- **Random Forests**: Given its durability and capacity to manage intricate feature interactions with minimal parameter modification, Random Forests may perform better in large datasets with a variety of features (such as brightness, color, and light curve properties).
- **SVMs**: In high-dimensional areas, support vector machines (SVMs) may perform better if the data is better organized and clearly distinguishes between different types of supernovae based on specific criteria. When there is a clear but complex class boundary, they are especially potent.

Because of its stability and capacity to manage intricate feature interactions, Random Forests may be more useful for large, noisy datasets with plenty of features. However, SVMs may be a better option for datasets when the data is in high dimensions and the classes are well-separated by a hyperplane, as they are excellent at identifying distinct decision boundaries even in intricate, high-dimensional environments.

5.4 *Enhancing Radio Astronomy Using Machine Learning*

The incorporation of machine learning techniques in radio astronomy has resulted in a significant improvement. These approaches are used to analyze complex data from radio telescopes, which frequently include signals from a variety of astrophysical sources. Machine learning algorithms may successfully separate these signals, find trends, and improve the identification of phenomena like pulsars, fast radio bursts (FRBs), and even possible extraterrestrial communications. Unsupervised learning techniques, such as clustering, have been used to group comparable signals, making it easier to locate and investigate novel sources of radio emission. Furthermore, AI approaches are utilized to calibrate and clean radio data, which increases the overall quality and dependability of observations [14].

From radio astronomy's inception to the present day of next-generation surveys, the graph in Fig. 5 shows the exponential growth in the number of radio sources found by major surveys over the decades. Interestingly, every notable increase in detections is correlated with developments in the technology of radio telescopes and the use of increasingly complex data analysis methods, such as machine learning. As seen, early surveys found only a few sources, including those by Grote Reber and the Mills Cross in Australia. However, the number of detected sources has risen into the millions with the introduction of surveys such as NVSS, FIRST, and more recent ones like EMU and VLASS. The enhanced sensitivity and resolution of contemporary telescopes as well as the incorporation of machine learning algorithms, which enable the more effective processing and interpretation of the complex data these equipment gather, are both responsible for this growth.

A key element in this evolution has been machine learning, which allows astronomers to manage massive data sets, remove noise, and detect weak signals that would have gone undetected with more conventional techniques. Finding new astronomical phenomena like Fast Radio Bursts (FRBs) and possibly even indications of extraterrestrial intelligence has been made possible through the application of unsupervised learning to categorize and group these signals. As the graph visually illustrates, in conclusion, the development of machine learning is a critical component of radio astronomy and will likely speed future discoveries and expand our knowledge of the cosmos.

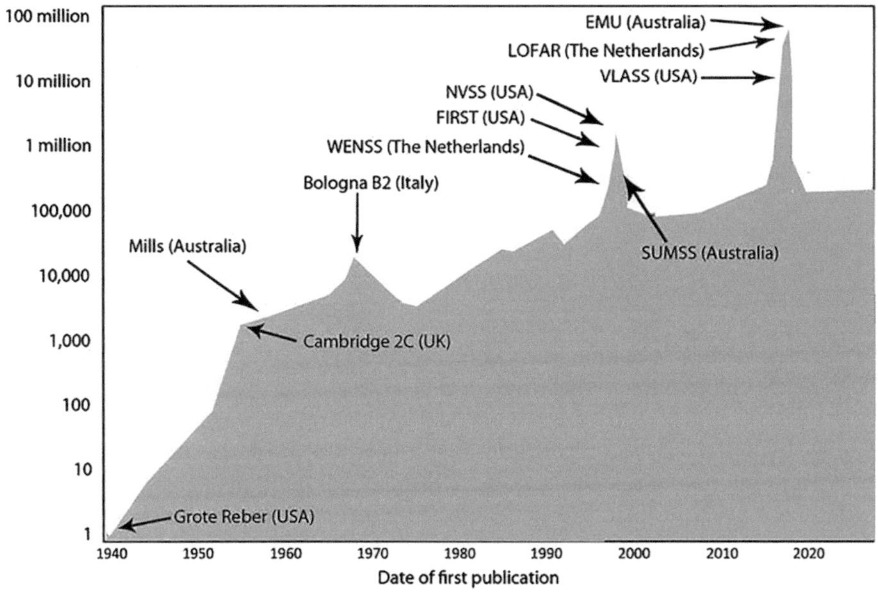

Fig. 5 Advances in surveys and machine learning techniques are driving an exponential increase in the detection of radio sources in radio astronomy (*Credit* Cosmosmagazine)

6 Future Directions and Challenges

6.1 Advances in AI Algorithms for Astrophysics

The future of astrophysics is set to be changed by ongoing advances in AI algorithms. Current artificial intelligence strategies, incl. deep learning and convolutional neural networks (CNNs), have already had a substantial influence; however, upcoming algorithms promise much larger advances. For example, generative adversarial networks (GANs) and reinforcement learning are being investigated for their ability to imitate celestial phenomena and optimize observational tactics. GANs can produce high-fidelity synthetic astronomical images to help train AI models on unusual events or to supplement existing datasets [15]. Reinforcement learning algorithms can schedule telescope observations to acquire the most scientifically valuable data [16]. These developments will not only improve our ability to detect and analyze astronomical objects, but will also aid in the solution of complicated cosmological and astrophysical problems. Transfer learning, which enables AI models trained on one piece of astronomical data to be transferred to different but related tasks, is another exciting development in the field. This is especially helpful in astronomy, where there may not be as many labeled datasets available. It allows models to use information from other fields to perform better on new tasks. In astrophysics, Explainable AI (XAI) is also gaining popularity as a means of improving the transparency and interpretability of AI models. Understanding how complicated models make decisions is crucial, particularly in high-stakes domains like the categorisation of cosmic occurrences or the forecasting of astronomical catastrophes. In order to find novel astrophysical phenomena that have not yet been classified, Unsupervised learning techniques like autoencoders are being improved to find patterns in huge datasets without the necessity for labeled data. When taken as a whole, these developments in AI algorithms are improving our ability to observe things and opening doors to new findings and understandings of the universe's most puzzling mysteries.

6.2 Integration of AI and Robotic Telescopes

The combination of AI with robotic telescopes offers a substantial advancement in astronomy study. Artificial intelligence (AI)-driven systems are able to make judgments in real time using data from observations and operate telescopes autonomously. This makes it possible to respond quickly to ephemeral events like rapid radio bursts, gamma-ray bursts, and supernovae. For example, the Zwicky Transient Facility (ZTF) employs machine learning algorithms to analyze data in real time, triggering additional observations when potential transients are recognised [17]. The combination of AI and robotic telescopes also makes large-scale sky surveys possible, as AI can pre-process and analyze data to find objects of interest, considerably decreasing astronomers' workloads and enhancing data gathering efficiency.

The graph in Fig. 6, shows the increasing trend in astronomy and astrophysics publications containing keywords such as "neural network" and "machine learning". There is a discernible and persistent growth in both categories starting about 2010, which is indicative of the increasing incorporation of these AI methods into astronomy study. The increase in publications is in line with developments in AI-driven robotic telescopes, which depend on these technologies to process large datasets, operate telescopes autonomously, and make judgements in real time while conducting observations. The graph illustrates how the scientific community has come to appreciate AI's benefits, especially when it comes to accurately and efficiently analyzing celestial occurrences. This trend is expected to pick up speed as additional robotic telescopes go online and as AI algorithms get better, which will result in even bigger contributions to theoretical advancements and observational capabilities.

In summary, the integration of artificial intelligence (AI) with robotic telescopes represents a significant breakthrough in astronomy, as evidenced by the expanding corpus of research that highlights the critical role of machine learning and neural networks in expanding our knowledge of the cosmos. This collaboration between observational astronomy and AI is expected to propel future discoveries at a never-before-seen rate.

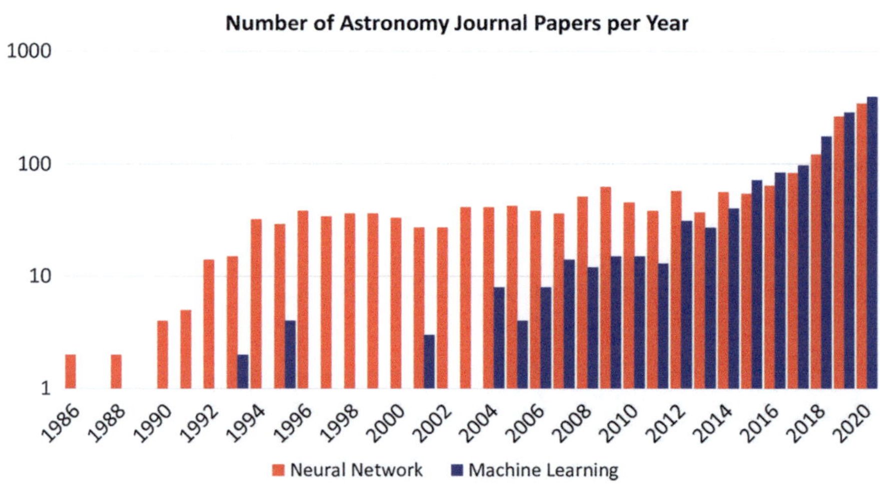

Fig. 6 Growing impact—the exponential rise in astronomy publications emphasizes the growing integration of machine learning and neural networks, which propels the development of robotic telescopes and AI-powered astronomical research. (*Image Credit* 'Machine Learning Applications in Astrophysics' by Soo, Al Shuaili, and Pathi.)

6.3 Future Prospects for AI in Space Exploration

AI's significance in space exploration is projected to grow significantly in the future years. AI systems are being developed to help with autonomous navigation, scientific data analysis, and decision-making in space missions. AI, for example, can be used to analyze data collected by rovers and landers on distant planets, identifying geological features and potential indicators of life with minimal human involvement. Furthermore, AI will play an important part in interstellar mission planning and execution, as well as trajectory optimisation and resource management on board. The employment of AI in conjunction with sophisticated robotics will allow for more intricate and long-duration missions, expanding humanity's reach farther into space and enabling for more thorough exploration of distant planetary bodies.

Unprecedented discoveries and advancements are anticipated as artificial intelligence (AI) technology advances and is integrated into space exploration. As mankind expands further into space, artificial intelligence (AI) will become increasingly important. It must be able to manage massive data quantities, adjust to unforeseen issues, and function independently in challenging environments. In the future, artificial intelligence (AI) might be used to power spacecraft that undertake extraterrestrial exploration missions, study the atmospheres of exoplanets, and perhaps help in the hunt for hidden intelligence. Combining artificial intelligence (AI) with other cutting-edge technologies will allow us to push the limits of space exploration, paving the way for future explorers to tackle the most challenging interstellar travel issues and opening up new research opportunities.

Acknowledgements The author(s) express their gratitude to the editors of this book for providing the opportunity to contribute this chapter. The two authors worked alone on this project; no other organisations or mentors contributed to its creation. Special thanks go to the book editors and reviewers for their insightful comments and guidance throughout the preparation of this manuscript.

Author(s) Contribution The main manuscript was written by Mr. Manjuleshwar Panda, who also carried out the literature review and conceptualised the chapter. Dr. Yogesh Chandra, the second author or mentor, contributed specific technical insights on AI approaches, with assistance from a few additional sources. The final draft of the chapter was examined and approved by all authors.

Funding No particular funding from public, commercial, or nonprofit organisations was obtained for this work. The research and writing were carried out independently by the author(s).

Data Availability The data used in this chapter were sourced from publicly available astronomical surveys and repositories, including NASA ADS and arXiv data archives.

Conflict of Interest The author(s) declare no conflicts of interest related to this work.

References

1. Tyson JA, LSST Science Collaboration (2018) The large synoptic survey telescope: unlocking the secrets of the universe. Nat Astron 2:10–14
2. Bloom JS, Richards JW (2012) Data mining and machine learning in time-domain discovery and classification. In: Proceedings of the IAU symposium, vol 285, pp 359–364
3. Domínguez Sánchez H, Huertas-Company M, Bernardi M, Tuccillo D, Fischer JL (2018) Improving galaxy morphologies for SDSS with deep learning. Mon Not R Astron Soc 476(3):3661–3676
4. Mitchell TM (1997) Machine learning. McGraw-Hill Education
5. Goodfellow I, Bengio Y, Courville A (2016) Deep learning. MIT Press
6. LeCun Y, Bengio Y, Hinton G (2015) Deep learning. Nature 521(7553):436–444
7. Abazajian KN, and associates (2009) The seventh data release of the Sloan digital sky survey. Astrophys J Suppl Ser 182(2):543
8. Starck J-L, Murtagh F, Bijaoui A (1998) Image processing and data analysis: the multiscale approach. Cambridge University Press
9. Howell SB (2006) Handbook of CCD astronomy. Cambridge University Press
10. Redmon J et al (2016) You only look once: unified, real-time object detection. In: Proceedings of the IEEE conference on computer vision and pattern recognition, pp 779–788
11. He K et al (2017) Mask R-CNN. In: IEEE international conference on computer vision proceedings, pp 2961–2969
12. Shallue CJ, Vanderburg A (2018) Identifying exoplanets with deep learning: a five-planet resonant chain around Kepler-80 and an eighth planet around Kepler-90. Astron J 155(2):94
13. Huertas-Company, M. and associates (2019) A deep learning approach to characterize stellar populations in galaxy images. Astron Astrophys 622:A27
14. Connor L et al (2018) Detection and localization of fast radio bursts using machine learning. Mon Not R Astron Soc 476(1):186
15. Mustapha W et al (2020) A generative adversarial network for astronomical image simulation. Astron Astrophys 635:A133
16. Lin H et al (2018) Reinforcement learning in astronomy: real-time strategy optimization for observational scheduling. Astrophys J Suppl Ser 237(1):24
17. Graham MJ et al (2019) The Zwicky transient facility: science objectives. Publ Astron Soc Pac 131(1001):078001

AI-Driven Insights: Revolutionizing Satellite Imagery Applications

Raj Kishor Verma, Indrajeet Kumar, and Sumit Kumar

Abstract Artificial intelligence has revolutionized applications of satellites in imagery by producing information previously impossible to acquire and operational efficiency on a wide spectrum of businesses. With a focus on crisis management, agriculture, urban development, and the monitoring of ecology, the technology addresses new applications in intelligent technology with satellite images. Tremendous volumes of data can be processed and analyzed speedily and accurately through a combination of AI technology and satellite snapping pictures. Image recognition was done using human judgment, which was cumbersome and limited. However, all of these have changed with the rise of deep learning and different machine learning approaches that will allow automatic object recognition and classification in satellite images. The usage of AI algorithms based on data set patterns may enable researchers to trace problem areas and predict future trends. Such information is vital to effective conservation plans and resource management. Emergency response teams look at enhanced images from satellites during severe weather events like wildfires, storms, or landslides for very critical information. The quick analysis of data allows responding agencies to get a grip on the extent of destruction faster and disburse supplies better. As another example, AI algorithms may compare before-and-after pictures to help compute damages and figures out which areas need the most urgent attention. In other words, it saves time while also pushing forward efforts at response overall.

R. K. Verma (✉)
Department of Computer Science and Engineering (Data Science), ABES Institute of Technology, Ghaziabad, India
e-mail: rvrajverma77@gmail.com; raj.24scse3010016@galgotiasuniversity.ac.in

School of Computer Science and Engineering, Galgotias University, Greater Noida, India

I. Kumar
School of Engineering and Technology, Birla Global University, Bhubaneswar, Odisha, India

S. Kumar
Department of IT, ABES Institute of Technology, Ghaziabad, India
e-mail: sumit.kumar@abesit.in

S. Chand Kumain et al. (eds.), *AI and ML Techniques in Image Processing and Object Detection*, https://doi.org/10.1007/978-981-96-7445-9_5

Keywords Artificial Intelligence (AI) · Satellite imagery · Machine learning · Data analysis · Real-time monitoring

1 Introduction

The potential of satellite imagery has been rapidly being pushed forward by AI-driven insights enabling previously unthinkable capabilities across numerous sectors. The proliferation of satellite data is only going to increase—driven by technological advancements in the satellite sector and a rise in open-source satellites—the use of Artificial Intelligence (AI) and Machine Learning (ML) has become critical for allowing us to retrieve insights from this ocean of information. AI in Satellite Imagery Satellite images can cover large areas therefore AI technologies increase the ability to analyze them and their objects, such as land use monitoring, and spectral analysis of environmental changes. Algorithms within AI can analyze complicated datasets that would be far too much for human analysts. The effectiveness of this information is incredibly important in fields such as agriculture, disaster management urban planning or climate monitoring where decisions need to be taken quickly but with correct information. With companies like Planet and Black Sky developing processing methods that enable AI to analyze satellite images straight off the press, the utilization of satellite data for predictive analytics has advanced significantly. By studying species through satellite pictures, artificial intelligence (AI) aids in keeping track of wildlife for environmental protection. Furthermore, by looking at trends in land use and environmental changes, AI will predict any possible crises in geopolitical situations. This change is AI-driven. It uses both advanced machine learning algorithms as well as new ways of processing data to interpret the complex datasets that are beamed to us daily from thousands of satellites. The efficiency of handling large volumes of satellite data is greatly improved by AI. Traditional manual analysis techniques are often laborious and subject to human error. With AI algorithms automating the procedure of extracting valuable information from pictures, one can quickly assess how the environment has evolved, and what the impact of land use and crisis response would be. For example, previous data can be used to train machine learning models with similarities so that they will categorize items, discover patterns, and even predict future changes [1]. AI is especially good at deciphering multispectral images, which gather data at multiple wavelengths. This capability helps to explain many things. For example, water quality, soil moisture content, and vegetation health. Looking at the spectra with AI brings essentials for management of resources and agriculture, and monitoring of the environment. By using in-depth spectral analysis to identify stress factors, crop health and harvest forecasts can be evaluated with AI-driven tools. AI's integration with the study of satellite data allows for insights that are nearly instant. Satellite companies like Black Sky are using AI to process pictures right as they take place, providing fast intel which can contribute to decisions for the average citizen and major corporate holdings alike [2].

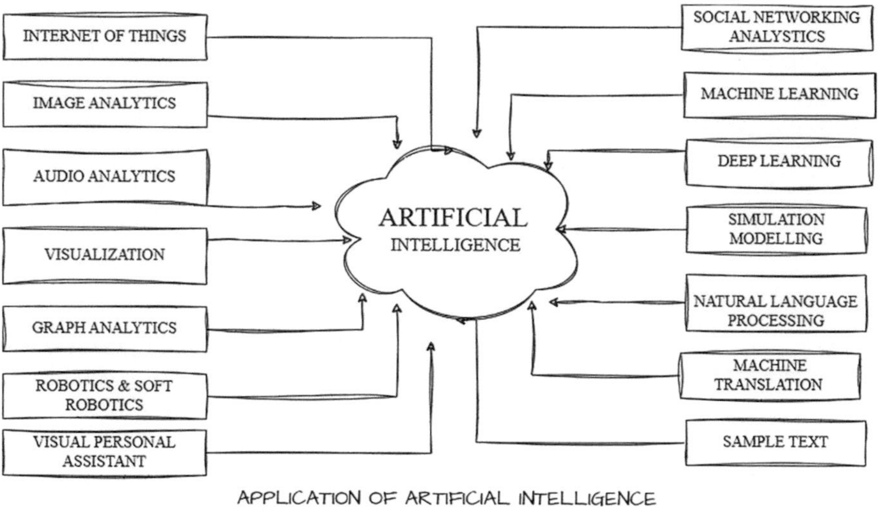

Fig. 1 Application of artificial intelligence

1.1 Artificial Intelligence (AI)

Machine-like and artificial as a transformative technology, intelligence typically performs tasks that are the normal province of human brains. In the field of AR, this idea takes shape on a variety of levels–from automation and individual services to predictive analysis below we introduce several forms or "channels" used by The sine qua non of a successful future for humanity is thus to make AI work for us. Meanwhile, AI is an immensely powerful vehicle for improving productivity qualitatively better choices about the future, and entry into all manner of inventions often across a widespread in industry. The crux of AI's future potential lies now in leading it responsibly. As AI continues to mature and conquer one problem after another, As AI becomes more closely integrated into our lives, many new uses will again be occasioned for it (Fig. 1).

1.2 Satellite Imagery

The technology has been successfully developed for use in a wide variety of industries. We shall see that this is especially the case in agriculture, environmental monitoring, urban planning and disaster management. Following this overview of satellite imagery and its public significance today, the next sections illustrate how various sorts of satellite data can tell us about society. Urban High-resolution cities for urban planning Cities use satellite imagery for the planning of urban land and infrastructure. These images help urban planners see land-use patterns, and determine the scope of urban sprawl and the method of construction necessary so citizens can live and work more efficiently [3]. Real-time With satellite imagery, data concerning earthquake areas and hurricane times at disaster locations during natural disasters can be provided. It assists in appraising the extent of injury suffered as well as conducting emergency relief work along with subsequent recovery efforts. Scientists are particularly interested in Snow flying machines equipped with instruments and radar-sounding equipment that cruise the globe through June to September recording snowfall levels, whether it is dry or wet (moisture content), its depth of accumulation as well number of days left until summer melts it away altogether [4]. Scientists Over 90% of the world's planets Satellite images are indispensable for climate research in many ways. They enable scientists to track temperatures, rainfall patterns, and sea levels worldwide in real-time; detect how wind blows around the globe can be visualized from space on a minute-by-minute basis. Monitoring Biodiversity Conservationists use satellite imagery to monitor wildlife habitats and changes in biodiversity. This information is of great use in carrying out their work to protect endangered species and improve the management of nature reserves [5] of National Infrastructure Satellite imagery is used for watching after infrastructure systems and facilities that are vital to economic development, such as roads, bridges, and railway lines. This comes from examining what needs to be fixed on the ground and maintaining the structural integrity of these facilities. 2.12 In telecommunications, satellite imagery is employed as a tool for network planning expansion. It provides information about all the geological features that may affect coverage including what conditions so engineers can input this data into computer programs like ArcViewGIS in order to generate computer-based maps [6].

The Technology Behind Satellite Imagery Sensors and Cameras Onboard satellites are equipped with advanced sensors that can gather different sorts of data. For example, an optical-sensitive sensor is used to capture visible light pictures; similarly, radar sensors in black and white capture the land seen from the sky. These sensors can map the earth no matter what kind of weather it is (rainy, snowy, foggy, dusty, etc.).

Data Processing the raw data taken from satellites is greatly processed so that it can be corrected: these are four things that are wrong with optical or microwave survey maps one must transform them. These processing steps include orthorectification (correcting geometric distortions) and radiometric correction (adjusting brightness levels).

Geographic Information Systems (GIS) GIS technology, integrated with satellite imagery, allows users to analyze spatial data effectively. It matches people together with various pieces of information obtained from satellites, distributing this with squaring on the map or constructing maps after measuring activity all over India's External Lands by adding satellite data where needed to fill in blanks. Artificial Intelligence (AI) Integration Integration with AI offers a higher-order power for the analysis of satellite imagery. Machine learning algorithms can help automate feature extraction and classify types of land use, extracting trends in consideration from historical data.

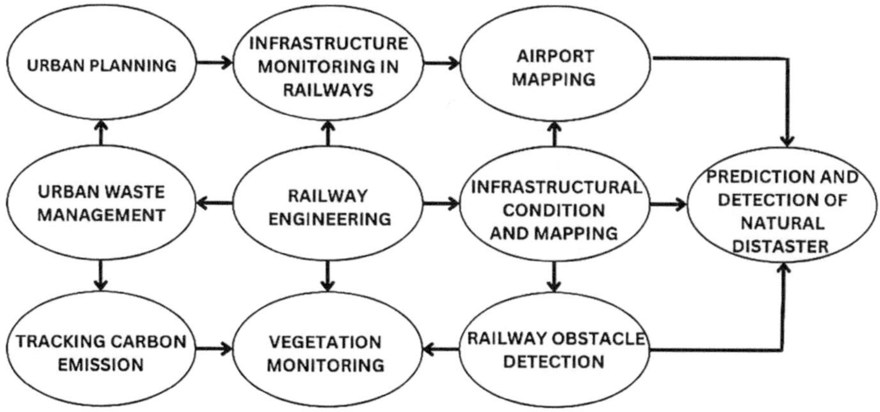

Fig. 2 Application of AI and satellite imagery

AI-driven applications will analyze large data quicker and more efficiently than those methods present based on traditional methods. Satellite imagery has become vital in a variety of fields: thanks to the information it offers decision-makers above ground. It extends from agriculture to disaster preparedness, making it a key resource all over the world. As the new technology of combining advanced AI 611—and specifically in integration with satellites under [7] makes its presence felt more broadly, satellite imagery will continue to find additional potential uses: and this also brings pleasant music for decision-makers to hear since it provides fresh opportunities in thinking about how we might best supervise and optimize our planet's requirements [7] (Fig. 2).

Satellites have been used for a long time to get a variety of information about the surface of the Earth, such as tracking surface vegetation, global weather patterns, ocean currents and temperatures and several others. The high resolution of images, coupled with the decrease in costs and greater availability made it increasingly possible for the public to use satellite images more fully. Satellite Imaging Technology has led to the development of hyper-spectral and multispectral sensors that can aid in finding objects, identifying materials and detecting processes. Satellite images are "digital images of the Earth's surface (or any other planet) compiled from spectral data collected by sensors carried in special-purpose satellites, readily available for all parts of the world from various commercial and government sources". In simple words, satellite imagery is images of the Earth (or any other planet) that is collected by imaging satellites operated by governments and businesses around the world [4].

- Use satellite imagery and artificial intelligence to monitor railway infrastructure. Railway tracks are one of the most fundamental and central components of railway systems. Even little structural deterioration might have devastating repercussions. Furthermore, the presence of any object or substance in the train's gauge may endanger the passengers' lives. Vegetation is one of the most significant invaders along the railway rails. It becomes considerably more perilous when vegetation develops and penetrates the train's gauge. Similarly, trees can touch or batter catenaries, slowing vehicles for safety concerns. Thus, satellite photos may be quite useful in monitoring and maintaining vegetation.

- Satellite technology Imagery and AI applied Urban Planning Rapid metro expansion and development have increased strain on ecosystems, particularly urban parks and green areas. Green areas are vital for improving urban environments and providing a quality of life for the urban population. Green places include lawns, public parks, gardens, streetscapes, woods, and so on. In this regard, technologies such as satellite imagery and AI/ML can assist urban developers and land [7] managers in observing and promoting decision-making for sustainable growth in dense urban environments, as well as preventing flooding in urban areas, by gathering high-resolution details concerning the urban area [8].
- Satellite imagery may give extensive analysis to locate significant changes in urban land cover and land use, allowing for frequent coverage and overlaying of different time frames to define ecologically safe and sustainable zones in every planned development area(s). Thus, satellite imaging, together with AI/ML [6, 9], can play an important role in assuring ongoing urban planning and growth [10].
- Natural disaster prediction and detection using satellite imagery and artificial intelligence (AI) Satellite pictures, combined with GIS maps, may provide a wealth of information for the evaluation, analysis, and monitoring of natural catastrophes such as hurricanes, tornadoes, volcanoes, earthquakes, and cyclone damage in local and big regions throughout the world. It may serve as a key tool and technology for monitoring and managing catastrophes, developing strategic planning models, and predicting and controlling natural disasters as they occur.
- Satellite Imagery and AI for [9, 10] the railroad Obstacle Detection. Obstacles on railways nearly invariably cause damage and accidents to trains since they are hard to avoid, posing a threat to passenger safety. It is critical to recognise them as early as feasible. Natural catastrophes and environmental factors can both provide challenges.
- Railway crossing monitoring can also be added here. Every year in France, there are an average of one hundred train-car crashes, with thirty persons killed and fifteen badly injured. Even though the majority of accidents are caused by human error, early discovery of a blocked crossing has a significant benefit. Obstacle detection would combine ongoing vehicle detection efforts with railway demands.
- Use satellite imagery and AI for infrastructure condition and mapping. Satellite imagery allows for the evaluation of various infrastructural situations. The high resolution of satellite photos has the ability to reveal fine features in the observed area. Satellite imagery, for example, enables the observation of broken railway tracks, broken/damaged roads, broken/damaged bridges, damaged catenary poles, damaged air bases, runways, and other such structures, allowing accidents to be avoided before they occur [11].
- Satellite Imagery and AI for Airport Mapping High-resolution satellite imagery, paired with AI, machine learning, and computer vision algorithms, may play

a critical role in the planning and construction of airport layout plans, navigational mapping, airport security, and aviation safety operations. 3D digital surface models and digital terrain models may be generated to offer data and details for the development of airport runways, terminals, layout design, airspace studies, obstacle surveys, facility mapping, taxiways, and other projects. Furthermore, using remote sensing satellite image data and GIS, airport planners and developers may gather all of the information required to improve traffic planning inside the airport structure. Furthermore, remote sensing satellite image data can be useful for recognising environmental changes, urban growth around the airport, changes in land use patterns, and vegetation behaviours in the airport region [12].

1.3 Machine Learning (ML)

Machine learning (ML) [9] refers to a branch of artificial intelligence which develops algorithms that allows computers to learn and make prediction from the data. Over the past few years, this technology has become extremely popular because it can find patterns from a very large amount of data that can be very difficult for human beings to distill manually. Here is all you want to know about machine learning—its types, techniques, applications, challenges and future evolution (Fig. 3).

Overview of Machine Learning

Automated learning refers to how computers, without being explicitly programmed for different given tasks, use algorithms to analyze data, learn from it, and make decisions or predictions Machines should generally get better with exposure to data, that is the goal [9, 11].

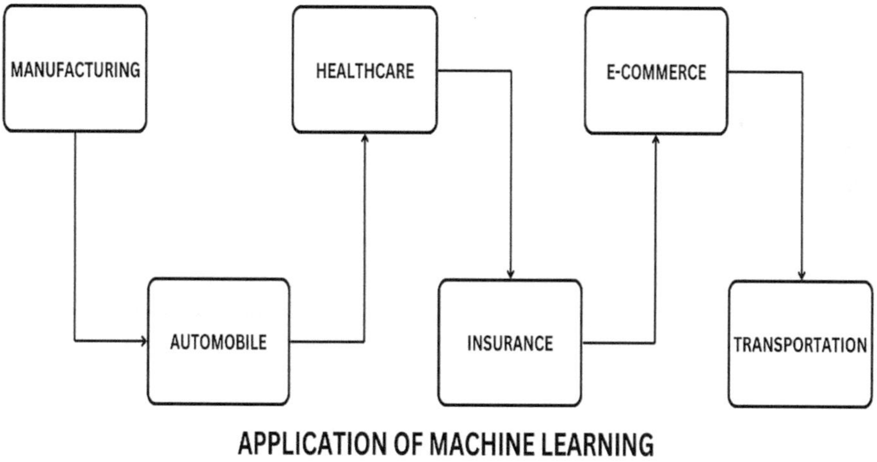

Fig. 3 Application of machine learning

1.3.1 Machine Learning Techniques

Machine Learning has become such a powerful and immensely popular sub-field under the umbrella term Artificial Intelligence that is works silently behind the scene and changes the way we people solve a problem or analyze the data. As an application of artificial intelligence, machine learning gives the system the ability to automatically learn and improve performance based on past experiences without being explicitly programmed, which is key to enabling computers to adapt to their foes and upgrade over time [1, 2]. An introduction to different types of learning.

1.3.2 Supervised Learning

Supervised learning You can think of supervised learning as an input and an output. Supervised Learning is to learn a function from input to output that can generalize and predict new and unseen data [13, p. 102]. An example of supervised learning: you show the algorithm a database of images of cats and dogs, along with the appropriate label (cat or dog) for each image, and it figures out which image is which.

1.3.3 On-Line Unsupervised Learning

These algorithms aim to reveal the hidden clusters, groups or associations in data when no target variable is being designated. Customer segmentation is one of the most common applications of unsupervised learning the task for the algorithm here is to discover groups of customers who are similar to each other in terms of purchasing behavior and/or other features without being given any labels in advance.

1.3.4 Data Analysis

Data analysis is the set of techniques used for observing, perforating out, transforming and building models of data in order to extract intelligent information and inform decision-making. It includes a variety of methods and techniques to help organizations make evidence-based decisions. Here are a ton of definitions, certainly in types, equipment and features conditions data oriented analysis. Data analysis is the science of collecting, inspecting, cleaning and transforming data with the goal of discovering useful information, drawing conclusions, and supporting decision-making. It is relevant to both numerical/quantitative data and categorical/qualitative data, and is important in many fields (e.g., business, healthcare, social sciences). Data-Driven Decision Making: Companies can draw insights of current trends and come up with strategically sound decisions using data analytics. Trend Identification: Combination of historical data analysis helps you identify trends will let businesses plan future strategies. Performance Improvement: Organizations can utilize data to

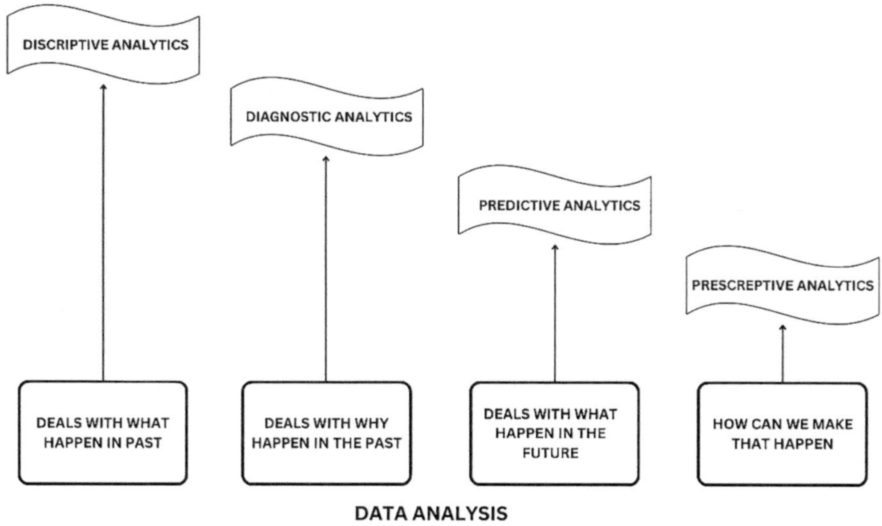

Fig. 4 Data analysis

evaluate their processes and identify improvement opportunities. Run of Data Analysis Different businesses make use of data analysis: Business Intelligence Health Processing the data of patients in hospitals or clinics improves treatment results and makes health institutions function better.

Data analysis: Businesses utilize data analysis to monitor key performance indicators and strategic choices.

Marketing: Improve marketing techniques and, communication with customers.

Social Sciences: In social science (research) the researchers analyze survey data or experimental data using statistical methods 1.

Data analysis is a vital process that makes sense of raw data by delivering meaningful insight in a wide variety of areas. From descriptive statistics to predictive modeling, organizations and methodologies for data analysis will also continue to develop, providing even more means for generating insights from data in this increasingly challenging space. The following are the most common techniques applied in data analysis; with definitions (Fig. 4).

1.4 Real-Time Monitoring

- This practice has been around for many years with real-time monitoring using real time performance and security, allowing organizations to quickly identify anomalies and problems in systems or events as they happen during the execution of various processes such as those found in manufacturing but also in healthcare where access to data needs being as fast possible, significantly affecting decision making and operational effectiveness. Real-time monitoring is the immediate collection and testing of data, followed by a prompt reporting of results for action-taking to solve problems quickly with minimal time lag between data collection and processing activities Sensors and sources of data basically are devices (network gadgets and application servers) Monitoring software consists of data collection and analysis apps, along with dashboards. User interfaces that interpret data insights into an understandable form. The advantages of live monitoring are that a merchandiser property owner can promptly catch difficulties and responsive steps can be taken to ease fallout. Better Control: Live surveillance is such that if it looks like someone has broken open at some time, or attempts to enter a secures area unauthorized alert will be instantly raised.

- Enhanced Performance: Because bottlenecks are instantly identified and can be solved before problems escalate, systems continuously function at high performance levels.
- Improved Decision Making: Real-time data is giving you the information as it is means strategic decisions developed based on current organizational circumstances and not based on historical points.
- Trend insights: This can reveal trends and patterns over a period of time which is beneficial for you to plan your next move.
- IT infrastructure monitoring—Real time has been used for IT Infrastructure Monitoring where organization monitor their IT environments, servers, networks that is it uses for monitoring the applications as well. So that it can be of better use to you and come in handy when required [6, 9, 10].
- Protection Tools: An Activity Based on Potential Security Under Network Safety This is a more, proactive and term that positions them for better cyber security posture.
- Application Performance Monitoring—Real-time monitoring of applications enables businesses to monitor their application performance metrics (response times, etc.) and troubleshoot issues that impact user experiences.
- Data Process Mining – In manufacturing environments, the processes are running in real-time stream to monitor performances of machines and production process for minimizing downtime and improving efficiency of operation.
- Health Care monitoring– In hospitals, real-time monitoring is used to track the patient vital signs and status of equipment that need prompt solutions.
- Financial Transactions Financial institutions have now properly embedded real-time monitoring systems that are able to detect fraud transactions.
- The Trouble with Real-Time Monitoring (Fig. 5).
- Huge quantities of data are produced daily. There might be duration in the analysis if you do not use its progression.
- Integration Issues: Integrated in real time monitoring tools [7, 14].
- False Positives—If too sensitive, alerts will get generated far more regular than it should be (potentially inflicting alarm fatigue amongst staff).
- Cost Sensitive: Full end to end real time monitoring solutions can be costly to use, both in software licensing and ongoing subscription maintenance.

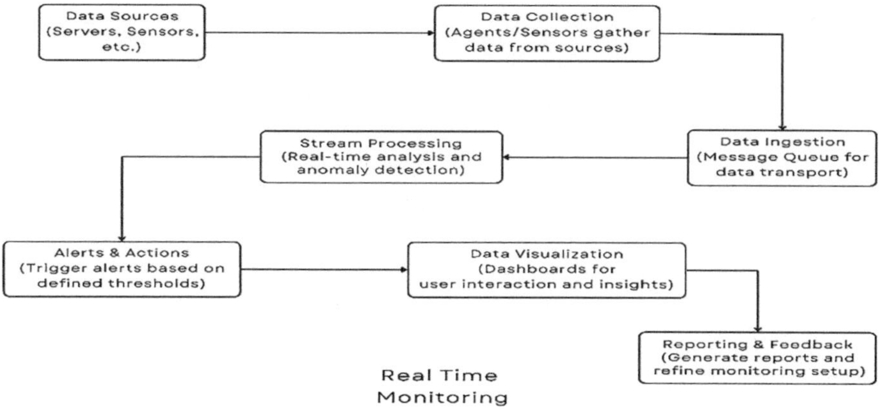

Fig. 5 Real-time monitoring

2 Literature Review

Get the Table 1 take a fast glance at AI based Insights and links to publications on the latest research, development of using satellite imagery: Revolutionizing Satellite Imagery Applications.

3 Proposed Methodology

Explanation of Each Component Reported Satellite Data Naked Image Input Unit Raw satellite imagery piled up into a file just like in when you first evaluated this item. Data Processing Phase Clean up the Raw Data With it goes unwanted things such as noise and shadows. This crucial step employs a wide range of AI techniques, including clustering, classification and outlier detection, to extract actionable patterns from the Data Layers. Real-Time [9, 11] Analysis With the data it's possible to make a real-time anatomization, enabling agencies to gain insight into and adapt rapidly changing conditions on land surfaces. Generating Insight This step converts the interpreted data into sharp insight actionable at every turn: in the fields of agriculture, community planning, and environmental monitoring. Results are viewed through dashboards and reports that help to advance market trade and trace probe detections closed open ended. Communication Stoned non-stop improvement in algorithms and processes goes beyond crystal methamphetamine by giving feedback on the perception created is part of this feedback loop, allowing users to be a part of its continuing enhancement (Fig. 6).

Table 1 Review of literature survey

S. N	Title	Authors	Publication date	Methodology
1	New satellite aims to show how AI advances earth observation	ESA	July 2, 2024	Discusses the sat-2 mission that utilizes AI for real-time processing of satellite imagery
2	AI-based system for satellite image analysis	Computer science journals	2023	Introduces an AI-based system for automating thematic information extraction from satellite images
3	How AI is turning satellite imagery into a window on the future	Defense one	June 2024	Explores how AI tools enhance satellite data analysis for predicting geopolitical events
4	Using AI and open source satellite imagery to address global problems	Omdena	May 17, 2022	Highlights projects combining AI with open-source satellite imagery for actionable insights
5	The role of AI in enhancing disaster response through satellite imagery	Various sources	2023	Reviews how AI-driven insights improve disaster response efforts using satellite imagery
6	Machine learning algorithms for satellite image classification	Various sources	2023	Overview of machine learning algorithms used for classifying features in satellite images
7	Deep learning and satellite remote sensing for biodiversity monitoring and conservation	Nathalie Pettorelli	2024	The nature crisis necessitates reliable and cost-effective tracking of biosphere changes
8	AI-driven approaches for real-time satellite data processing and analysis	Hafez Ahmad	2024	Focuses on AI methods applied to monitor environmental changes via satellite data
9	Integrating AI with satellite data for urban development planning	Jiadi Yin	2021	Analyzes how integrating AI with satellite data can facilitate urban planning
10	Real-time processing of satellite imagery using AI technologies	Hafez Ahmad	2024	Highlights advancements in real-time processing of satellite imagery through AI technologies

(continued)

Table 1 (continued)

S. N	Title	Authors	Publication date	Methodology
11	Generative models in satellite imagery analysis: future directions	Hadi Mansourifar	2022	Discusses the potential of generative models in enhancing satellite imagery analysis
12	AI and ethical accounting: navigating challenges and opportunities	Beatrice Oyinkansola Adelakun	2024	Identifies current challenges and explores future opportunities for AI applications in satellite imagery
13	AI-powered satellite imagery analysis: a review	Various sources	2023	Reviews various AI techniques applied to satellite imagery analysis, highlighting advancements and challenges
14	New satellite to show how AI advances earth observation	ASD news	July 2, 2024	Discusses the capabilities of ESA's Φsat-2 satellite in utilizing onboard AI for Earth observation
15	How AI is enhancing marine ecosystem monitoring via satellite imagery	Various sources	2023	Examines how AI applications can be utilized to monitor marine ecosystems using satellite imagery

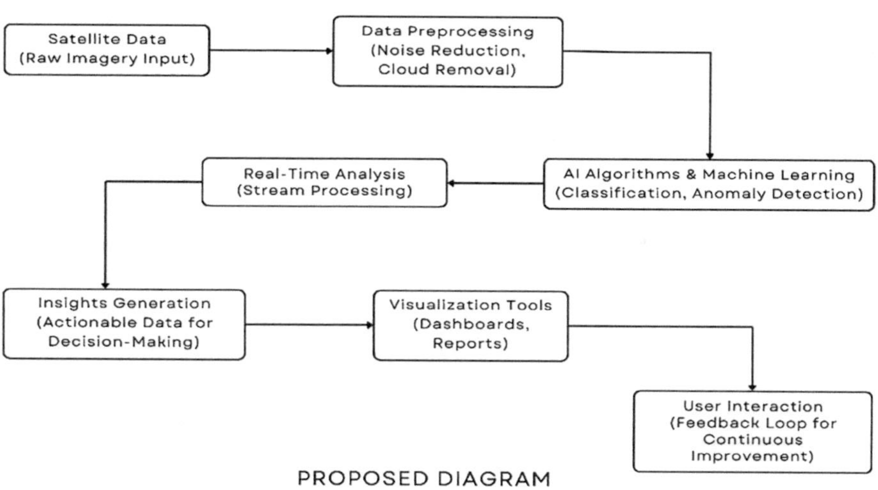

Fig. 6 Proposed diagram

4 Conclusion

An AI-based data-driven approach to satellite imaging services is revolutionizing several sectors such as agriculture, urban planning, and environmental monitoring. These algorithms have the potential to revolutionize the approach organizations take towards geospatial data, offering unprecedented analysis at scale and an ability to react to changing input in real time, promoting sustainability in production itself. This revolution enables specific crop management with resource destruction, while also enhancing disaster response and monitoring of urban infrastructure. Yet data quality, bespoke model design, and trust-building with stakeholders should still be prioritized. With AI technologies continually developing and more mainstream competitors entering the space, the possibilities for further disrupting industries with advanced imagery applications are only going to expand.

5 Future Scope

A cutting-edge AI approach using terabyte-dense high-resolution satellite data, which is growing exponentially, presents a myriad of challenging problems that are poised to change how Earth is perceived and understood. As this research direction crosses several fields, including computer vision, machine learning, geoscience, physics, big data exploration, and human–AI interactions in health, environment, and society, it will benefit both the AI community and those in other fields who will be the users of these future AI [7] models. In a fast-paced, data-driven world, current research and prospective thoughts as laid out herein can be utilized in setting and realizing future satellite data strategies not just to enhance human understanding of the Earth and its surrounding environment but also for the betterment of our future existence. The interaction between AI and remote sensing is not new, which dates back to more than 30 years ago. The digital exploitation of Earth observation big data is based on AI algorithms, including clustering, classification, feature-specific recognition, and change detection. The focus of these AI applications has been predominantly driven by 'what we can do' according to sensors and communication link hardware [6, 14] performance evolution. Especially over the past decade, AI has demonstrated its ability as an enabling technology to process petabyte-sized photonic global databases. However, this revived collaboration has been in a somewhat passive vein, propelled mostly by the requirements of business interests, such as mining, insurance, defense, and precision agriculture. More radical AI adoption within EO data analysis pipelines would now present an opportunity to turn the question around and lead with 'what we should do' for our future stake in the use of satellite assets and for the general benefit of our planet. It is now high time that the current momentum in the fields of AI and remote sensing continues to grow for the betterment of our existence.

However, progress in AI for satellite image analysis has been limited due to a lack of labeled data and the varying visual and geometric aspects of satellite images.

Nevertheless, satellite imagery is widely used in environmental, urban, and agricultural planning, military operations, and navigation assistance. We highlight the insights [9, 11] that AI-driven satellite imagery can provide in fields such as agriculture, environmental science, urban planning, and defense. We also recognize the unique challenges of satellite imagery and advocate for advanced visual learning methods to address them. We recognize that combining satellite imagery with AI is a potent tool for discovering valuable insights into historical and contemporary cultures, heritage sites, archaeological discoveries, and important national or international events. Despite the progress made by current AI-based satellite image analysis solutions, this area of research is still in its early stages and presents significant opportunities for future projects, from data collection to analysis to implementation. We believe that this partnership is likely to lead to a range of innovative future solutions [9–11], and we encourage the next generation of researchers, institutions, and businesses to join in and help shape the future world through both visual and descriptive perspectives. Emerging Technologies in Satellite Imagery Analysis.

6 Challenges

It helps manage those resources efficiently, discover local opportunities, and assist with everyday tasks. This is related to AI-enabled insights in Satellite Imagery Analytics and is observed, stressing the fact that combining AI and analytics is key to achieving better learning and ultimately more accurate insights from the data. Planet, whose constellation of tiny satellites capture daily photos of the planet, creates an astonishing data set. But the path from image pixels to information is interrupted by the very data being received from space—an overwhelming number of pixels that in many cases are impossible for humans, and even for machine learning algorithms, to process [9]. This paper examines how AI-derived insights can assist in deepening our understanding of our planet's emerging insights.

The state of the art in satellite image analytics.

These methods are often slow, and may not be adequate for the huge workloads that modern satellites produce every day. Challenges include: Cloud Cover: Atmospheric conditions can obstruct images and make analysis difficult [7].

Data Volume: With some satellites producing as much as 80 terabytes of imagery in a single day (a volume that is human-way-too-much to process), Complexity: There is a high level of dependency on data experts to understand the complicated data sets, reducing accessibility [14] for a lot of organization.

Satellite images have helped several businesses to deploy AI-driven solutions with success [6] Omdena Project: In AI+ Satellite Imagery Open Source Projects, Combining AI with open source satellite imagery, Omdena developed tools for enhancing image resolution and classification accuracy for a variety of use cases (achieving up to 99% accuracy in identifying land use patterns). Planet Labs, a leading satellite imagery company that leverages advanced AI models to quickly analyze satellite data in order to inform geopolitical strategies and disaster response efforts.

References

1. Omdena (2022) Using AI & open source satellite imagery to address global problems, 17 May 2022. https://www.omdena.com/blog/ai-satellite-imagery
2. AIT R, Siddha S (2024) AI-based system for satellite image analysis: Landuse and land cover classification. Int J Comput Artif Intell 5(1):09–14. https://doi.org/10.33545/27076571.2024.v5.i1a.75
3. Defense One (2024) How AI is turning satellite imagery into a window on the future, June 2024. https://www.defenseone.com/technology/2024/06/how-ai-turning-satellite-imagery-window-future/397520
4. Kishor K, Saxena N, Pandey D (eds) (2023) Cloud-based intelligent informative engineering for society 5.0, 1st edn. Chapman and Hall/CRC. https://doi.org/10.1201/9781003213895
5. Agrilinks (2023) Geospatial models: Unlocking the potential of satellite data through AI. https://agrilinks.org/post/geospatial-models-unlocking-potential-satellite-data-through-ai
6. Verma RK, Kishor K, Jha SK (2024) Big data analytics in bioinformatics and healthcare. In: Advances in bioinformatics and biomedical engineering book series, pp 25–43. https://doi.org/10.4018/979-8-3693-2426-4.ch002
7. Kishor K, Verma RK (2023) Cloud computing-based smart agriculture. In: Sharma A, Chanderwal N, Khan R (eds) Convergence of cloud computing, AI, and agricultural science (). IGI Global, pp 120–136. https://doi.org/10.4018/979-8-3693-0200-2.ch006
8. Kumar A, Verma RK, Rani R (2023) Edge cloud computing-based model for IoT. In: Kishor K, Saxena N, Pandey D (eds) Cloud-based intelligent informative engineering for society 5.0. Chapman and Hall/CRC, pp 123–140. https://doi.org/10.1201/9781003213895-7
9. Verma RK, Kishor K (2024) Innovative solutions. Chapman and Hall/CRC eBooks, pp 155–177. https://doi.org/10.1201/9781003489368-8
10. Verma RK, Mittal D, Sharma M, Jindal R (2024) Enhancing sentimental analysis using multimodal data. CINEFORUM 65(3):98–125. https://revistadecineforum.com/index.php/cf/article/view/74
11. Verma RK, Kishor K, Galletta A (2024) Federated learning shaping the future of smart city infrastructure. Chapman and Hall/CRC eBooks, pp 196–216. https://doi.org/10.1201/9781003489368-10
12. Mansourifar H, Moskovitz A, Klingensmith B, Mintas D, Simske SJ (2022) GAN-based satellite imaging: a survey on techniques and applications. IEEE Access 10:118123–118140. https://doi.org/10.1109/ACCESS.2022.3221123
13. Yin J, Dong J, Hamm NAS, Li Z, Wang J, Xing H, Fu P (2021) Integrating remote sensing and geospatial big data for urban land use mapping: a review. Int J Appl Earth Obs Geoinf 103:102514. https://doi.org/10.1016/j.jag.2021.102514
14. Verma RK (2024) Image processing applications in agriculture with the help of AI. In: Infrastructure possibilities and human-centered approaches with industry 5.0. IGI Global. https://doi.org/10.4018/979-8-3693-0782-3.ch010

Fundamentals of Deep Learning in Image Analysis and Object Detection

Mudit Mittal, Vivek Kumar, and Partha Sarkar

Abstract Over the past few years, deep learning has been a drastic change in how images are being analyzed with ever increasing accuracy particularly when it comes to tasks like image classification, image segmentation etc. This part elucidates the major principles and basic architectures of deep learning and focuses on its purpose to understand images. The first part of this chapter defines and explains neural networks and is particularly concerned with the specifics of deep learning as a trend of machine learning while fully or partially eliminating the need on any prior feature engineering. This chapter lays out the intricacies of training of Convolutional Neural Networks (CNNs), the most popular architecture for tasks associated with images. CNNs learn the spatial hierarchies and patterns of images making such networks critical for performing tasks like object detection, image recognition and segmentation. Important also is understanding what comprises CNNs described one of the chapter's sections including convolutional layers, pooling, fully connected layers and how purposes of those components assist in image processing effectively. Next along the lines of following developments in medical imaging technologies, what is covered in the chapter is also dedicated to other modern methods for biomedical images processing, such as U-Net and Fully Convolutional Networks (FCN), which are used for pixel-wise images segmentation. Certain special focus is directed on the use of these models for classification and segmentation of white blood cells, which is an important and difficult area of clinical diagnostics. The chapter provides key challenges that come with the application of a deep learning approach towards the image analysis, including but not limited to overfitting, data imbalance, and an absence of interpretability. Possible solutions for these challenges, especially in contexts with little labelled data will also be reviewed, including the use of dropout, data augmentation and transfer learning. In conclusion, the chapter delineates the

M. Mittal (✉) · V. Kumar
Department of CSE, THDC-IHET, New Tehri, Uttarakhand, India
e-mail: muditmittal@thdcihet.ac.in

V. Kumar
e-mail: vivek@thdcihet.ac.in

P. Sarkar
Department of IT, Sparsh Himalaya University, Dehradun, Uttarakhand, India

S. Chand Kumain et al. (eds.), *AI and ML Techniques in Image Processing and Object Detection*, https://doi.org/10.1007/978-981-96-7445-9_6

probable future developments in the field, which include attention mechanisms and self-supervised learning technologies which will take deep learning models for image analysis a step further than their current state. As a result of the defining of the key features of the deep learning application as well new approaches, this chapter seeks to create an understanding of the impact that deep learning is having on image analysis and imaging in general especially in areas such as biosciences and other medical applications.

1 Introduction to Deep Learning and Image Analysis

Definition: ***Deep learning is considered as a subset of artificial intelligence (AI) which trains and guide systems to read data exactly same as the human brain reads. The identification of sophisticated structures in images, words, audio, and additional data results from deep learning algorithms, producing accurate insights and predictions*** [1].

1.1 Overview of Deep Learning

Deep learning is an approach that revolutionized how images are analyzed to unprecedented accuracy and how machines can understand and interpret data in visual format. Embarking from this context, chapter introduces the basic principles of deep learning, which are related to the applications of image analysis. We shall discuss how DL models, especially CNNs, are designed, trained, and used to extract meaningful information from images [2].

Deep Learning relates to the universe of Machine Learning. It refers to the technical aspect of feature extraction by multi-layered neural networks from unprocessed input through classification of learning architectures. Also, Deep learning has achieved phenomenal results in activities such as natural language processing & image and voice recognition. Deep learning has advanced fields like computer vision and artificial intelligence by automatically learning features.

1.2 Importance and Applications

Deep learning has exhibited some remarkable advantages in superseding the old-fashioned image processing techniques that confine the extraction of feature extraction completely to automation, and mainly become relevant in obtaining very high performance with huge quantity of data.

Through deep learning's power of enabling automatic feature extraction image analysis progressed dramatically, hence minimizing the need for manual involvement. It excels in object identification, segmentation, and classification. Deep learning enhances diagnostic accuracy in domains like medical imaging by detecting minute patterns in pictures, increasing speed, precision, and decision-making.

Applications of image analysis include Object detection, Segmentation, Classification, and Anomaly detection-all are very pertinent to a broad range of areas, such as autonomous cars, satellite imaging, medical imaging, and surveillance systems.

1.3 Historical Context

The history of deep learning originate from the 1950s and 1960s, with the development of artificial neural networks. Over the decades, researchers have continued to refine and improve deep learning models, leading to major breakthroughs in the 2000s and 2010s. Traditional image processing techniques relied heavily on manual feature engineering, where domain experts handcrafted features to be used by machine learning models. Rather, autonomous deep learning architectures are capable of detecting discriminative features in raw image data, thereby needing minimal manual feature extraction [3].

1.4 Fundamentals of Deep Learning, Machine Learning and Neural Networks

Understanding basic machine learning and artificial neural networks will be important first in making clear understanding about the domain of deep learning. It is a subdivision of AI including specific models are given based on criteria or objectives and such models are allowed to deduce their own conclusions with no explicit programming. Simplistically, ML feeds on data-driven learning from examples to dig out underlying patterns and generalize from unseen data.

In the broadest meaning, machine learning is the overall term under which many techniques and algorithms fall that allow systems the ability to learn from data with minimal explicit programming. Supervised, unsupervised and reinforcement learning are the major ideas in machine learning.

Designed largely based on the human brain structure; artificial neural networks are composed of neurons, which are interrelated nodes, which can send signals to one another. The more data the network is exposed to, the better it learns patterns and generates predictions.

Neural networks are driven by physiological neural networks in the individual's human brain, are a fundamental framework for machine learning. These interconnected nodes act like neurons and pass signals between each other, adjusting the strength of the interconnections in response to the data.

2 Core Concept of Deep Learning

Machine learning has a division called deep learning that concentrates on advanced neural networks with extensive layers. Each layer extracts progressively extensive features from raw data which is given as input. It replicates the capacity of human minds to acquire knowledge through experience and evolve, making it practical for applications like image authentication, natural language processing-(NLP) and speech identification.

Some prime deep learning models are as follows.

2.1 Artificial Neural Networks (ANNs)

Artificial Neural networks normally considered as key requisite of deep learning approach, which draws inspiration from the naturally aging neurons seen in the individual's human brain [4]. Neurons comprises neural networks, which are linked nodes that exchange impulses with one another. This network adapts to execute specific tasks, including classification of images, by modifying the connections to the underlying neurons through a training process.

Information is processed and passed through these networks of linked nodes, or neurons. Each neuron performs a stimulus function on the weighted total of all its inputs before sending the output to the layer below. The non-linear relationship among the input data and target outcomes may be learned through neural network by the effective use of the activation functions. The specific network architecture, including the number and size of the hidden layers, determines the types of functions that the neural network can represent.

2.2 Deep Neural Networks (DNNs)

Deep Neural Networks are the foundation of deep learning. Many hidden layers are involved in the designing of such deep neural networks. DNNs have changed the working environment of AI in the present modern era. The quality of artificial intelligence projects has significantly improved due to recent research developments in deep learning and neural networks [5].

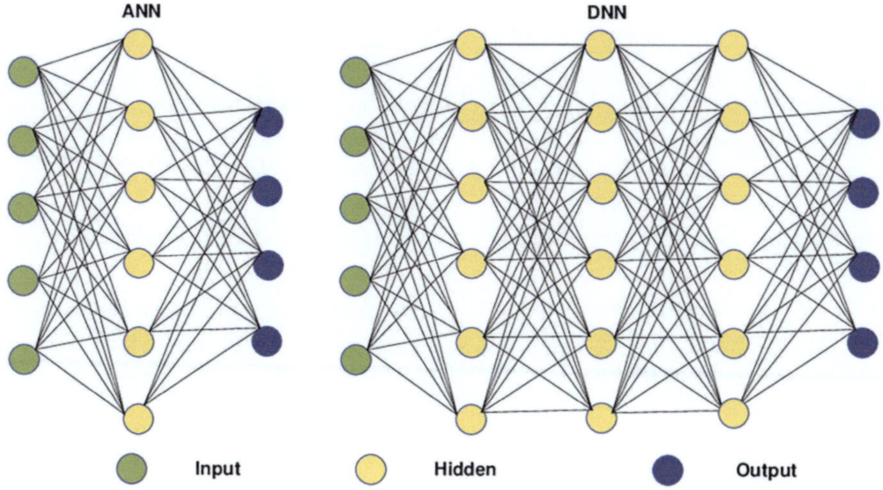

Fig. 1 Structure of ANN and DNN [6]

DNNs use deep architectures to learn complicated and reliable data representations. Empirical evidence of its expressivity and robustness in training algorithms over thousands of classes on the difficult ImageNet classification challenge demonstrates the ability to acquire sophisticated object representations without the requirement for hand-designed features.

These deep neural networks assist programmers in producing better and more long-lasting outputs. Because of this, they are even taking the place of numerous traditional machine learning methods.

DNNs are being incorporated as an important component in many cyber-physical systems, such as the vision system of a self-driving car to better recognize pedestrians, vehicles, and road signs (Fig. 1).

2.3 Supervised and Unsupervised Learning

These are two fundamental prototypes in the field of machine learning. Both have their own extraordinary characteristics and dedicated applications. Moreover, Supervised learning is considered as a category of machine learning in which training for a system is involved on a particular data-set with labelled inputs and corresponding outputs, with the goal of learning a function that can accurately map the inputs to the outputs. This approach is commonly used in classification and regression problems, where the algorithm aims to forecast a category-based output variable which depends on the input features.

In unsupervised learning, the system's training is done on data-set without any labelled outputs, with the goal of discovering the underlying structures and patterns

Table 1 Difference among supervised and unsupervised learning

Supervised learning	Unsupervised learning
Receives labelled data as input	Receives unlabelled data as input
Possesses a feedback system	Avoids a feedback system
Training datasets are used to classify data	Offers the provided data attributes in order to classify it
Separated into classification and regression	Separated into association and clustering
It is used for forecasting	Analysis is done by using this
It contains algorithms like decision trees, logistic regressions, support vector machine	It contains algorithms like k-means clustering, a priori algorithm and hierarchical clustering
Its classes are known in numbers	Classes are unknown in terms of numbers
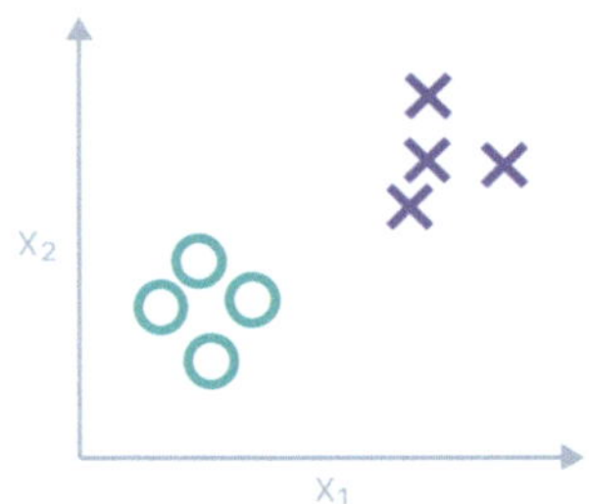	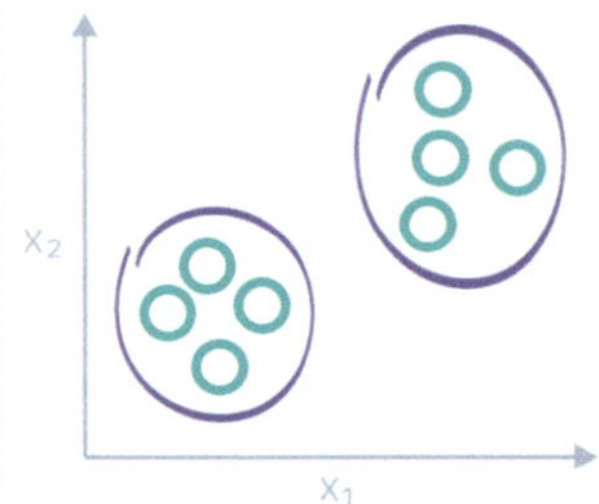

present in the data. Some Unsupervised learning algorithms, like clustering and dimensionality reduction, are often used for tasks like data exploration, anomaly detection, and feature extraction.

One key distinction between the two approaches is the level of human supervision involved. Supervised learning requires a significant amount of human effort to label the training data, whereas unsupervised learning can uncover insights from the data without the need for prior labelling. Table 1 shows the significant difference between both learnings.

3 Convolutional Neural Networks (CNNs) for Image Analysis

There are so many effective ones among neural networks. Among these, the most effective and stable topologies of a neural network for the application of image analysis is the convolutional neural network.

For effectively capturing the spatial and local properties of pictures, convolutional neural networks are trained. This makes them appropriate for using wide variety of areas and applications, including object identification, analysis of medical pictures, image segmentation and many more.

Images can be thought of as high-dimensional data, and with millions of pixels, fully connected layers within the standard neural network would have an impractically large number of parameters to model such high-dimensional input. This problem is addressed in Convolutional Neural Networks through the exploitation of local spatial structure in image [7].

Therefore, CNNs apply the convolutional filter to the local regions of the image, which allows efficient weight sharing as well as spatial invariance; this makes them ideal to use in tasks like medial image analysis.

The three prime components of CNN are (i) Set of Convolutional Layers, (ii) Pooling Layers, and (iii) Fully Connected Layers.

3.1 Set of Convolutional Layers

Convolutional layers serve as the foundation for CNNs. To determine the dot product among the filter and local input areas, the convolution technique slid a filter or kernel across the input picture.

The final feature map size is specified when filters are applied with padding and stride during convolution. By building up a large number of convolutional layers, a network may be trained to recognize hierarchical visual representations. Filters are used to capture certain patterns like edges, textures, or forms.

3.2 Pooling Layers

CNNs frequently employ pooling layers to minimize computational complexity and decrease the spatial dimensionality of the feature maps. The most popular pooling procedures are the average pooling and maximum pooling. Average pooling is used to compute the average value within the pool area and maximum pooling maintains the largest value within the pool area.

- **Max Pooling**: This Pooling chooses a local region's maximum value. This helps capture the most prominent features in the local receptive field.
- **Average Pooling**: computes the average of values within a local region. This helps in smoother feature maps and preserving overall information.

Pooling helps the network generalize better by reducing sensitivity to small translations in the input image.

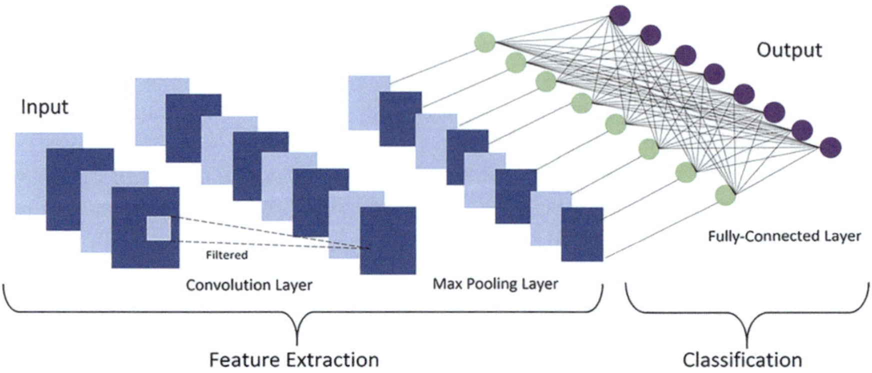

Fig. 2 Design of CNN [8]

3.3 *Fully Connected Layers*

The combination of fully connected layers is contemplated as the ready and final output module of any CNN architecture. The feature maps are usually routed via fully connected layers after being flattened by many convolutional and pooling layers. These layers combine the extracted features into a high-level representation, which is used to make final predictions (e.g., class labels). The output layer usually employs a softmax activation function for classification tasks.

These techniques can help deep learning models to adapt the unmatched and classified properties of medical imaging data, such as differences in image modalities, anatomical structures, and disease patterns.

Convolutional layers, as shown in the Fig. 2, apply many sets of learnable filters on the input medical picture. It makes the network detect low-level features, which include edges, textures, and shapes. Then the network applies pooling layers for feature maps downsample, thereby decrementing the spatial dimensions as well as the computational complexity related to the network. Finally, in the end, the set of fully connected layers of such network aggregate all learned features to make a prediction such as classifying the input image under some particular class.

Convolutional Neural Networks (CNNs) have significantly relevant for the analysis area for medical imaging, where the superior performances are achieved in the tumor detection, organ segmentation, and disease diagnosis tasks.

3.4 *Receptive Fields and Strides*

In Convolution Neural Networks, the concept of the receptive field represents the region in the input image that a feature in a certain layer is looking at. As we dive much deeper into such complex network, the receptive fields of features in higher layers become larger, allowing them to capture more global and contextual information.

The stride parameter determines how the convolution filter is applied to an input image. The meaning of 1 stride of is that filter is moves with one pixel at a time, whereas a larger stride for example 2, skips every alternate pixel, decreasing the spatial dimensions of the output feature map.

These architectural decisions in CNNs allow them to effectively extract and hierarchically compose visual features from image data, making them powerful tools for a variety of medical imaging tasks.

3.5 Spatial Invariance and Weight Sharing

The prime advantage of using CNNs is its adaptability to learn spatially invariant features. This means that any specific feature such as like a texture or an edge, will be detected anyway of its exact location present in an input image.

This is enabled by the weight sharing mechanism in convolutional layers, where the same set of learnable weights (the filter) are applied across the entire input. As a result, CNNs require fewer parameters compared to fully connected networks, making them easier to train and more effective, especially for high-dimensional inputs like medical images.

4 Related Technologies for Image Analysis

Although primarily used networks for image analysis are convolutional neural networks, there are many more deep architectures from which to choose to address specific challenges or requirements. Deep Learning Techniques give comprehensive coverage for image analysis.

Some of the notable deep architectures include:

4.1 Residual Networks (ResNet)

These were derived to counter the vanishing gradient problem when making very deep networks. They introduce skip connections that allow the network to skip layers, enabling very deep architectures. A residual connection is used in a ResNet, where the network will learn a residual function, improving performance and the reliability of training [9].

4.2 Generative Adversarial Networks (GANs)

Artificial, novel information may be generated using GANs in order to duplicate the training set. This allows for the creation of tasks like picture synthesis and style transfer. GANs are composed of two main networks, first is a generator network which creates a new output and second is a discriminator network which evaluates if an output is real or synthetic [10].

4.3 Autoencoders

Autoencoders usually considered as unsupervised neural networks, aiming to learn a compressed data representation after training the network for reconstruction its own inputs. Autoencoders have gained prominence for their ability to learn useful features and compressed representations without any labelled data.

4.4 Transformer Models

Transformer models represent a unique architecture in neural networks that depend solely on attention mechanisms to identify long-distance connections in data, especially on natural language processing and medical image analysis.

Selecting a deep learning architecture is completely based on some specific application, the availability of training data-set and availability of the computational resources at hand. In the following sections, we will focus mainly on convolutional neural networks, which are considered as the leading deep learning architecture for image analysis.

5 Training Deep Learning Models for Image Analysis

A prime challenge in utilizing deep learning to medical image analysis is the less availability of large datasets which are labelled enough.

To resolve this limitation, transfer learning is extensively used by the researchers, which included adjusting deep learning models that had previously been trained on big datasets like ImageNet for the medical imaging job at hand.

It does turn out to be the case that transfer learning allows it to extend the execution of highly competitive deep learning models on a medical imaging task by merely using relatively small datasets to leverage the general feature representations learnt on natural images.

Other domain adaptation techniques are also developed to bridge the source and target domains more closely, such as natural images and medical images.

5.1 Data Preparation

As the quality of input data significantly influenced the execution and performance of the model, it is crucial to preprocess the medical imaging data rigorously before feeding it into the deep learning pipeline. This includes Image Resizing all images to a consistent spatial resolution and normalizing pixel intensities. And also addressing any missing data or artifacts that may exist in the medical images.

Data Augmentation approaches are widely adapted for further expansion of the limited training data. Such approaches helps in synthetically transformation of original training samples, through various operations such as rotation, flipping, scaling, adding noise, and more.

5.2 Data Augmentation

In image analysis, training models requires large amounts of labelled data, which is often scarce or expensive to obtain. By transforming existent photos with flips, translations, noise, rotations, and other operations, data augmentation methods are frequently used to fictitiously enhance the training datasets. Hence by this, overfitting can be reduced and improvement in the ability of the model for generalization of new sets of data.

5.3 Regularization Methods

Regularization methods are used to prevent over-adjusting if models performs good in training data but have low performance in visible data. Some common regularization methods in deep learning include:

- **Dropout**: Randomly eliminates neurons during training, causing the network to become less dependent on any one neuron and hence more resilient.
- **L2 Regularization (Weight Decay)**: The dimensions of model weights inform the incorporation of a penalty within the loss function.
- **Batch Normalization**: Normalizes the inputs to every layer, improving training stability and performance.

5.4 Transfer Learning

By using Transfer Learning technique, already trained model, normally trained in a large dataset (such as ImageNet), is perfect for specific tasks on a small data-sets. This method is particularly useful for medical image analysis, where the data marked is limited. By using the knowledge acquired from a general task, the model can achieve high accuracy for a specific domain task with fewer training examples [11].

6 Key Challenges in Deep Learning for Image Analysis

Medical Imaging is a very vast field. To make potential analysis of medical imaging, deep learning has proven quite promising. Nevertheless, there are unique challenges and opportunities that must be addressed. Many challenges have still existed in the field of image analysis, despite its tremendous advances [12].

6.1 Data Scarcity

Many image analysis tasks, especially in specialized domains like medical imaging, suffer from limited labelled data. Techniques like transfer learning and unsupervised learning are actively researched to address this issue.

6.2 Limited Availability of Large Datasets

The limited availability of large and annotated data-sets in the medical domain, is one of the prime challenges. To resolve this, researchers have explored techniques like as data augmentation, weakly supervised learning, and generative models to generate synthetic data and leverage unlabelled data.

6.3 The Interpretability

The Deep learning techniques sometimes reflected as "black boxes," which makes these model gruelling to grasp the process of electing the decisions. Efforts to improve model understanding and comprehensibility are critical, particularly in giant sector applications like as healthcare.

6.4 Computational Complexity

Large datasets for deep learning model training demand a substantial amount of processing power. Researchers are seeking for ways to enhance the model's performance through various substantial strategies like distillation, model pruning and quantization.

6.5 Variability in Medical Data

Another key challenge is the inherent variability and uncertainty in medical data, which can be caused by factors such as image acquisition, patient physiology, and disease progression. Deep Learning models must be capable enough of managing this uncertainty and providing reliable and robust predictions. Despite these challenges, the opportunities for deep learning around medical image analysis are vast. Deep learning is poised to transform medical image analysis by delivering unprecedented gains in accuracy, efficiency, and consistency, which are essential for improved patient outcomes.

Furthermore, the incorporation of deep learning approaches with other emerging technologies, like medical sensors, genomics, and electronic healthcare records, could enable more comprehensive and personalized healthcare solutions.

The explainability and interpretability of deep learning approaches furnish a major challenge in the medical field, as clinicians need to have a clear understanding of the procedure for making decisions. To overcome this, researchers have developed techniques like attention-based models, saliency maps, and explainable AI to provide more interpretable and transparent deep learning models.

7 Applications in Image Analysis and Object Detection

Segmentation of images is a computer vision technique that split a digital picture into distinct groups of pixels, known as segments of the image, to aid object recognition and other tasks. Image segmentation allows for quicker and more advanced image processing by breaking down an image's complex visual data into properly formed segments.

7.1 Image Classification

It is one of the subtasks of the computer vision domain where an algorithm or a model predicts the label of an image. In very simple terms, it constitutes a large part of the solution to a huge issue in Machine Learning and Deep Learning especially in fields like medical imaging, autonomous driving and facial recognition. Here's an overview of how it works and its key components:

Key Steps in Image Classification:

1. **Input Image**: The process begins with an image (e.g., a photo of a dog or a cat).
2. **Feature Extraction**: The model identifies distinguishing features in the image (e.g., edges, textures, shapes).
3. **Prediction**: The model uses these features to predict the class of the image (e.g., "dog" or "cat").
4. **Evaluation**: Accuracy is evaluated using metrics like precision, recall, F1-score, etc.

7.2 Semantic Segmentation

Semantic segmentation allocates a class label to all pixels of the image. This task is essential in applications like medical imaging, where it is crucial to accurately segment regions of interest, like tumors or organs. CNNs have shown quite effectiveness in semantic segmentation since they are well suited to extract spatial information and hierarchical features.

Well-known structures for semantic segmentation are:

- **U-Net**: It is a fully connected convolutional network intended to the segmentation of biological images, consisting of an encoder-decoder structure by skipping the connections to preserve spatial information.
- **SegNet**: It is a CNN-based architecture which uses an decoder and encoder based framework for image segmentation at the pixel level.

7.3 Object Detection

Multiple object localization and identification are part of object detection in a picture. When it comes to object detection, the model must provide bounding boxes and class labels for everything that is present, in contrast to classification of the image, by predicting a single label for the whole picture [13].

Common object detection architectures include:

- **YOLO (You Only Look Once)**: It is an actual object identification method which creates a grid out of the picture and forecasts the class probabilities and bounding boxes for all the cells.
- **Faster R-CNN**: Using a second CNN for classification and refinement, a region proposal network produces potential object regions in the first step of the two-stage process.

7.4 *Anomaly Detection*

Anomaly detection refers to finding cases in visual data that show a pronounced difference from usual patterns. These anomalies could be anything from unusual objects, structures, or patterns in images that don't conform to the expected or "normal" distribution. Its use is significantly important and useful specially in the applications where abnormality detection is much complex and condemning like medical imaging, industrial inspection, and security.

Types of Anomalies:

- **Point Anomalies**: Particular cases that are deviated from other cases.
- **Contextual Anomalies**: Points of data that can be thought of as strange in some respect, be it time or space.
- **Collective Anomalies**: A group of instances that together form an abnormal pattern but individually may not seem anomalous.

8 Future Directions

Even though deep learning has advanced medical picture analysis significantly, much more work and creativity remains. Some key future directions include [14]:

- Developing more robust and generalizable models that can handle variations in data and maintain reliable performance.
- Incorporating domain knowledge and prior information to improve model interpretability and make predictions more clinically meaningful.

8.1 *Self-Supervised and Unsupervised Learning*

To tackle the problem of lack of data, researchers are exploring the concept of self-supervised and unsupervised learning. These concepts could be adapted so that the valuable representations could be learned based on the unlabeled data, which then could be fine-tuned using smaller datasets with labels.

Some promising directions include:

- **Contrastive learning**: Obtaining representational knowledge through the comparison of positive and negative data sample pairs.
- **Generative models**: To create artificial medical pictures for data augmentation, generative adversarial networks or variational autoencoders are used.

8.2 Integration with Some Other Existing Technologies

The deep learning's fusion with other cutting-edge and modern technologies, like medical sensors, genomics, and electronic health records, could enable more comprehensive and personalized healthcare solutions. For example, combining deep learning-based image analysis with patient-specific genomic information could lead to more accurate disease diagnosis and personalized treatment plans.

8.3 Federated and Distributed Learning

To address the challenge of limited data availability, researchers are exploring federated and distributed learning approaches. These techniques allow models to be trained on data distributed across multiple sites or institutions, without the need to centralize the data, which is often not feasible in the medical domain due to privacy and regulatory concerns.

9 Conclusion

The chapter presents a brief summary of the main concepts and methods that forms the basis of deep learning for image analysis, from its revolutionary design to its mathematical foundations, like U-Net and CNNs. Algorithm, architectural, and computational resource improvements are continually refining deep learning and making even greater boundaries for what is attainable in image analysis.

Convolutional Neural Networks along with another deep learning architectures provide indeed impressive performance in medical imaging like image segmentation, lesion detection and disease diagnosis. Medical imaging introduces some unique challenges related to data availability, requires explainability and has inherent variability in the data-a set that requires special techniques and deep learning advances.

Deep learning methods has undoubtedly achieved a massive success in application area of medical imaging, ranging from radiological image analysis to pathology slide analysis, and many more.

Deep learning has made tremendous innovations whereby machines can, to-date, classify, segment, and even better detect objects in images better than at any other time in the past. Because neural networks are at the heart of deep learning, they prevail as an encouraging tool for spatial information capture from images, the closest one being convolutional neural networks.

Future of deep learning promises further integration with other fields such as genomics and electronic health records, potentially transforming personalized healthcare for the betterment of medical imaging. However, addressing the inherent variability in medical data and improving the transparency of deep learning system will be critical to gaining trust and broader adoption in clinical practice. Despite these challenges, deep learning holds immense capabilities to boost diagnostic reliability and patient outcomes.

References

1. LeCun Y et al (2015) Deep learning. Nature 521(7553):436–444
2. Ruiz-del-Solar et al (2018) A survey on deep learning methods for robot vision. arXiv:1803.10862
3. Ahmed SF et al (2023) Unveiling the frontiers of deep learning: innovations shaping diverse domains. arXiv:2309.02712
4. LeCun Y et al (2015) Deep learning. Nat Portfolio 521(7553):436–444
5. Szegedy C et al (2013) Deep neural networks for object detection. In: Proceedings of conference on neural information processing systems, pp 2553–2561, December 2013
6. Aslam S et al (2021) A survey on deep learning methods for power load and renewable energy forecasting in smart microgrids. Renew Sustain Energy Rev 144:110992
7. O'Shea K et al (2015) An introduction to convolutional neural networks. arXiv:1511.08458
8. Hashemi A et al (2023) Multibody dynamics and control using machine learning. Multibody Syst Dyn 58:397–431
9. Abdi M et al (2016) Multi-residual networks: improving the speed and accuracy of residual networks. arXiv:1609.05672
10. Creswell A et al (2018) Generative adversarial networks: an overview. IEEE Signal Process Mag 35(1):53–65
11. Raghu M et al (2019) Transfusion: understanding transfer learning for medical imaging. In: Proceedings of 33rd international conference on neural information processing system at Cornell University, pp 3347–3357, January 2019
12. Razzak I et al (2017) Deep learning for medical image processing: overview, challenges and future. Cornell University, pp 323–350, January 2017
13. Zhao ZQ et al (2019) Object detection with deep learning: a review. IEEE Trans Neural Netw Learn Syst 30(11):3212–3232
14. Zhou SK et al (2021) A review of deep learning in medical imaging: imaging traits, technology trends, case studies with progress highlights, and future promises. Proc IEEE 109(5):820–838

Comparative Analysis of Different Disparity Estimation Architectures on Aerial Datasets

Ishan Narayan and Shashi Poddar

Abstract With the advent of aerial image datasets, dense stereo matching has gained tremendous progress. This work analyses dense stereo correspondence analysis on aerial images using different techniques. Traditional methods, optimization-based methods, and learning-based methods have been implemented and compared here for aerial images. For traditional methods, the architecture of Stereo SGBM is chosen while using different cost functions to get an understanding of their performance on aerial datasets. Analysis of most of the methods in standard datasets has shown good performance; however, in the case of the aerial datasets, not much benchmarking is available. Quantitative and qualitative analysis of different disparity estimation techniques has been carried out over the stereo aerial datasets. Using existing pre-trained models, recent learning-based architectures have also been tested on stereo pairs along with different cost functions in SGBM. The evaluation of obtained depth maps has been carried out using different quantitative metrics such as MSE, BMP, and SSIM. Through the analysis, the author summarizes the performances of different methods and provides a way forward for disparity estimation techniques in the future.

Keywords Stereo images · Depth estimation · Semi-global block matching · Unmanned aerial vehicle · Learning based methods

1 Introduction

Unmanned aerial vehicles (UAVs) for mapping, surveillance, and remote sensing make use of sensor along with vision-based methods to generate accurate representation of the aerial view. One of the major challenges in ensuring their autonomous navigation and safe operation is their ability to select a landing site autonomously in any unknown area [1]. Key issues include ensuring safety by avoiding obstacles,

I. Narayan · S. Poddar (✉)
Academy of Scientific and Innovative Research, Ghaziabad, India
e-mail: shashipoddar@csio.res.in

CSIR—Central Scientific Instruments Organization, Chandigarh, India

S. Chand Kumain et al. (eds.), *AI and ML Techniques in Image Processing and Object Detection*, https://doi.org/10.1007/978-981-96-7445-9_7

maintaining efficiency through minimal human intervention, and providing flexibility to adapt to various terrains. Recent advancements in multi-sensor fusion and deep learning algorithms enhance UAV landing capabilities, enabling safer operations across complex environments. With simultaneous localization and mapping frameworks, navigation in a GPS-denied environment and selecting a suitable landing site have become very effective these days. Recent research highlights the development of a multi-sensor framework that enables real-time target localization, ensuring precise landing even in challenging terrains like grasslands and slopes [2].

However, most of the proposed works on autonomous landing of UAVs still rely on predetermined knowledge of the landing zone. Typically, a map of the selected area is utilized to train or identify potential landing sites. In the case of isolated locations or unknown scenarios, UAVs may need to land autonomously on different kinds of surfaces, which may be rocky, flat, or slanted. By integrating data from multiple sources, such as topographic maps and satellite imagery, machine learning algorithms can enhance the accuracy of identifying potential landing zones. Various parameters, such as depth, inclination, steepness, and flatness, are evaluated to determine the safety of a potential landing site. A flat surface is essential to minimize the risk of tipping or rolling during landing manoeuvres. Additionally, steepness and inclination are critical for assessing the overall safety of a landing site, as they directly impact the stability and control of the UAV during descent and touchdown. A dataset featuring a variety of surface inclinations was proposed by [3]. It consists of different surfaces at different inclination angles. This dataset can be used to test and refine new methods for assessing surface characteristics. Aerial images often include trees, buildings, rooftops, and diverse terrain types, all of which can complicate depth estimation. As a result, depth maps derived from aerial imagery can vary significantly in quality and accuracy.

Determining flatness, inclination, and steepness from stereo images requires a depth map of stereo pair, there exist several techniques as discussed in literature section. However, when applying some of the dense disparity estimation schemes to the UAV stereo images, the depth estimation architecture does not work with the same accuracy as that for a traditional set-up where the camera views the objects in front of it. Since most of the existing algorithms are optimized for indoor environments and do not generalize well enough to aerial images. Despite their effectiveness, stereo disparity estimation faces several challenges in the case of aerial platforms that can be attributed to several factors, such as occlusions, texture-less regions, environment variability, and the presence of different regions in the same image frame [4].

Although several benchmarks have been developed in the past to compare the performance of disparity schemes, the images from a UAV or aerial platform are not included in these challenges. Therefore, it is necessary to benchmark the performance of different disparity estimation schemes specifically for aerial images. These aerial images have specific challenges, such as low image resolution, low textured area for feature matching, and varied depth distribution. The low texture areas in an aerial image can be attributed to the fact that aerial images usually capture the top view of any region or area, which could be a building, vegetation, or areas that are either flat or inappropriate for feature matching.

Among several available architectures, SGBM-based architectures, optimization-based architecture, and learning-based architecture are benchmarked together. The SGBM-based algorithms are popularly used for different applications and is known for balancing computational efficiency and depth estimation accuracy. The optimization-based schemes provide relatively better depth estimates at the cost of higher computational time and are not fit for real-time application. Optimization-based methods were used as they provided visually appealing results comparable to learning-based methods. The learning-based techniques that use deep learning architecture have been compared for their performance on aerial datasets. As per the author's best knowledge, this kind of comparison has not been done specifically for aerial images, and with the rising popularity of autonomous UAVs for various applications, this study will help devise the future path of depth estimation architectures for images taken by UAV cameras. The overall article is divided into four sections of which Sect. 2 provides a brief overview of different disparity estimation schemes in the literature, Sect. 3 details the SGBM, optimization, and deep learning-based algorithms selected here for experimentation, Sect. 4 analyses and benchmarks these algorithms on two publicly available datasets the WHU stereo dataset [5] and Mid-Air dataset; and finally Sect. 5 concludes the paper.

2 Literature Review

Dense depth map estimation is a computer vision approach that aims at obtaining the depth of an image point given images from two rectified cameras with a known baseline distance. Although several disparity estimation algorithms have been proposed in the literature, it is still a challenging task to handle occlusions, texture-less regions, and discontinuities in the images. The images captured from aerial platforms face challenges like large disparity search space, bigger occlusions, and varied distribution. It is thus necessary to study the aerial stereo images holistically and benchmark the performance of different classes of disparity estimation algorithms on them. Dense depth map estimation algorithms using traditional methods is classified into three approaches global, local, and semi-global and these methods differ in terms of the cost aggregation methodology used [6]. With the rise in the usage of deep learning approaches for depth map estimation, these can also be classified as traditional or learning-based methods.

2.1 Traditional Depth Estimation Techniques

Traditionally, the four key processes involved in the depth estimation process are: matching cost calculation which calculates the pixel wise difference on the epipolar plane, then cost aggregation for obtaining the correct disparity pixel followed refinement. Various cost functions such as absolute differences (AD), the squared difference (SD), normalized cross-correlation, mutual information (MI), and several non-parametric methods and methods that use window-based approaches like rank and census transform. Other cost functions include AD-Census [7] in which SAD and Census transform are used together, or a combination of SAD and gradient function [8] have been explored in the literature. In some of the techniques, the image is divided into textured and texture-less regions for different cost functions, and some of the techniques incorporate larger kernel size and smaller kernels near occlusions. Several of these techniques use local adaptive windows and other cost functions, which are not discussed here further for brevity purposes.

Global methods like belief propagation [9] use Markov random fields to approximate minimum cost labels in the energy function. In [10], the authors proposed an efficient belief propagation algorithm that uses a hierarchical approach to reduce the computation time and memory usage. Another approach involving local alpha expansions based on an MRF model with continuous label space which applies different alpha labels according to the index. Dynamic programming-based algorithms independently perform scan line-based optimization for all scan lines in the image. In it, the authors used the RANSAC-based method to detect occlusions and assign labels accordingly. Among the global approaches, the patch match [11] based approach that uses iterative propagation and refinement from neighbouring pixels to obtain disparity value is also very popular.

Some other works use Minimum spanning tree (MST) [12], Super-pixel based clustering (SLIC) [12], and iterative clustering algorithms to obtain disparity maps and have better performance than local methods.

2.2 Learning Based Depth Estimation Techniques

Most of the global stereo disparity methods discussed above have an inherent disadvantage of being computationally intensive. There is a significant amount of research for the use of deep learning architectures (Poggi et al. 2021) to solve the bottlenecks in stereo disparity estimation problem. Training CNN for matching cost between image patches was initially introduced by Zbontar and LeCun [13] and currently, a significant number of end-to-end stereo matching networks have been developed, offering substantial advancements in disparity estimation by jointly learning all stages of disparity computation. This holistic approach has demonstrated improved performance and accuracy. DispNet [14] was among the first end-to end network. Several methods leveraging 2D convolution have also shown promising results like GwcNet [15], Stereo Transformer (STTR) [8], HITNet [16], AANet [17]. GwcNet introduced an enhanced cost volume representation through a group wise correlation volume. New methods also make use of concatenation-based feature volume though 4D cost volume along with 3D CNN to aggregate features like GC-Net [18] and PSMNet [19]. GC-net uses a 2D convolution neural network to obtain dense features.

CRE Stereo uses a neural network with an adaptive correlation module to register locations in multi-scale feature space. It uses a stacked cascade architecture and works by down sampling the image pair before constructing the image pyramid. Recent methods like RAFT-Stereo [20] first uses a encoder decoder based architecture to obtain an initial disparity map at different scale then different scaled versions of the depth map are used in refinement module. It uses R-CNN for cost aggregation and works by down sampling to lower resolution, thus saving memory and computation resources [21]. CNN based techniques display a significant accuracy boost over the previous approaches.

However, using 3D CNNs presents significant challenges due to their computational complexity and high memory requirements. While most of these approaches work well with datasets like KITTI [22], Middlebury, KITTI, NYU Depth etc. The model trained on any synthetic dataset usually does not generalize well in new scenarios. It is necessary to note that a model pre trained on a synthetic dataset cannot easily be applied to a real scene dataset due to the heterogeneous data sources. There are several methods that have used fine tuning for transfer learning. By training on diverse datasets, learning-based methods can generalize well and infer disparities even in challenging texture less regions.

3 Disparity Estimation Architecture

The disparity estimation architecture traditionally works by computing costs, performing aggregation along multiple directions and is followed by refinement. These schemes can be either local, or global, or semi-global in its approach. Local algorithms select stereo disparity for every pixel while global approaches use an energy function that is minimized over the complete data it has. In this section, the traditional semi-global block matching with its three cost variants, optimization-based Patch Match and learning based techniques considered in this article for experimentation is discussed theoretically.

3.1 Semi Global Block Matching

Semi-Global Block Matching (SGBM) is a stereo vision algorithm that offers low computational time and good accuracy in both static and dynamic situations. The primary attribute for a SGBM approach is the cost function which is a measure of disparity estimation. The cost is calculated for each pixel at all potential disparity levels and yields an initial disparity map with some errors. This cost value is stored in a row major sequence and the neighboring pixel cost being the adjacent row is represented as a cost volume. As observed, aggregation can be run in loop to aggregate for each path separately [23]. OpenCV provides the functionality to choose between 4, 8, and 16 paths, and the 4 or 8—path is generally used to maintain computation

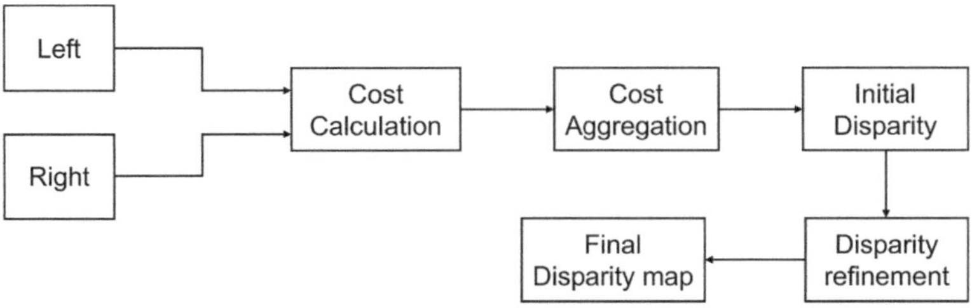

Fig. 1 Essential steps for the traditional SGBM pipeline

time within limit. Cost aggregation function E(d) is defined as:

$$E(d) = \sum_{p} (C(p, D_p) + \sum_{q \in N_p} P_1 T[|D_p - D_q| = 1] + \sum_{q \in N_p} P_2 T[|D_p - D_q| > 1]). \tag{1}$$

In Eq. 1, P_1 is the constant penalty applied when disparity value changes by 1, and P_2 is used when disparity value change is greater than 1. The aggregated cost for a pixel p is calculated by aggregating costs from all directions equally. The cost $L_r(p, d)$ along the path traversed in direction of r can be defined as

$$\begin{aligned} L_r(p, d) = C(p, d) + min(L_r(p - r, d), L_r(p - r, d - 1) + P_1, L_r(p - r, d + 1) \\ + P_1, \underset{i}{min} L_r(p - r, i) + P_2) - \underset{k}{min} L_r(p - r, k) \end{aligned} \tag{2}$$

The costs L_r is then summed over all the 8 paths as $S(p, d) = \sum_r L_r(p, d)$. In Eq. (2), $L_r(p - r, d)$ is aggregated cost when disparity of the last pixel in the path is d. $L_r(p - r, d - 1)$ is the aggregated value when disparity of the last pixel in the path is d − 1, min $(L_r(p - r, i))$ is the minimum value of all costs from the previous pixels. P_1 and P_2 2 are input parameters that can be tuned as per the dataset. The working flow of Stereo matching for disparity can be seen in (Fig. 1). The three different cost functions used here for comparing SGBM approach is described here briefly.

3.1.1 Birchfield–Tomasi Dissimilarity

Birchfield–Tomasi (BT) dissimilarity measure is a technique for gauging pixel similarity and is focused on determining the absolute variations in pixel intensities within a small area. A window is placed with its center at the left pixel, and a horizontal scan line is made through the corresponding right pixel to calculate the BT dissimilarity measure between that pixel and the pixels along the epipolar line in right image. The BT dissimilarity measure is used for stereo matching in OpenCV SGBM function and is defined as:

$$d_l(x_l, x_r) = \min_{x_r-\frac{1}{2}\le x\le x_r+\frac{1}{2}} \left|I_l(x_l) - \hat{I}_r(x)\right|$$

$$d_r(x_l, x_r) = \min_{x_l-\frac{1}{2}\le x\le x_l+\frac{1}{2}} |\hat{I}_l(x) - I_r(x_r)| \tag{3}$$

Here, x_l and x_r are the corresponding left and right pixel intensity values for left and right images I_l and I_r, respectively and $\hat{I}_l$ and $\hat{I}_r$ are the linear interpolation of the left and right images I_l and Ir, respectively. Instead of matching pixel to pixel, BT interpolates the pixel intensities and searches for half more pixels to get a more accurate match.

3.1.2 Sum of Absolute Difference

The sum of absolute difference (SAD) is a pixel-based matching cost function that works by computing the difference between the pixel intensity in a local window. SAD is simple and computationally efficient but is sensitive to brightness and contrast changes, making it less effective in certain scenarios.

3.1.3 AD-Census

This cost function is a combination of absolute difference and census transform, wherein the census transform is a binary descriptor that encodes the spatial relationships between pixels in a local window. From left and right image kernels along epipolar line is calculated using the Hamming distance between their census transforms. The hamming distance is a measure of difference of bit positions in two binary strings and is represented as C_{census}. For every pixel, both the costs are calculated and added after normalizing them using their respective constants, resulting in one cumulative cost defined as:

$$C_{AD}(p, d) = \frac{1}{3} \sum_{i=R,G,B} \left|I_i^{Left}(p) - I_i^{Right}(pd)\right| \tag{4}$$

$$C(\text{p}, d) = \rho(C_{census}(\text{p}, d), \lambda_{census}) + \rho(C_{AD}(p, d), \lambda_{AD}) \tag{5}$$

where $\rho(c, \lambda)$ is a cost function dependent on c and λ, described as:

$$\rho(c, \lambda) = 1 - exp(-\frac{c}{\lambda}) \tag{6}$$

Here, $C_{AD}(p, d)$ is the cost function describing the absolute difference, C_{census} is the cost value obtained through census transform, and C is the cumulative cost value.

3.2 Optimization-Based Patch Match

Among the several optimization-based techniques for dense depth map estimation, a patch match-based scheme is selected here for analysis and comparison. The patch match algorithm finds the nearest matching patches between two images and uses random initialization [11], iterative propagation, and search for nearest neighbor-based estimation. This process computes correlations on neighboring pixels and sends its cost to the next matching point in an iterative way. The random initialization aspect estimates random disparity on an image to get a normalized plane and merge it to a plane equation later. This is followed by an iterative propagation which tries to reach a global minimum. In this paper, single-iteration results are obtained, and the number of iterations can be increased to obtain slightly better results.

3.3 Deep Learning Based Methods

Learning-based methods have shown promising results in stereo disparity estimation and are an evolving approach that needs further improvement as well. Supervised methods for stereo disparity need labelled data in which ground truth disparity maps are provided for training the network. These methods use convolutional neural networks (CNNs). During training, the networks are optimized using a loss function that is used to improve predictions by measuring the difference between the predicted result and the ground truth disparity maps. In this work, Cascaded Recurrent Network with Adaptive Correlation, HITNet and RAFT Stereo has been tested and their pre-trained models have been used for experimentation.

3.3.1 CRE Stereo

In this technique, a two shared-weight feature extraction network is applied on both the images that outputs a feature pyramid. This feature pyramid is used for estimating an initial depth at different scales in the 3-stage cascaded recurrent network. This method starts with 1/16 of the input image resolution. The first level in cascade used the original version of the input stereo while the other levels are fed up-sampled version for initialization. The output from each stage is fed to correlation layer for matching ambiguities in case of non-rectified stereo pairs. Group-wise correlation for cost volume is used simultaneously and preserves the details in high-resolution input.

3.3.2 HITNet

HITNet framework has been designed for depth estimation and addresses the challenge of achieving real-time performance while maintaining accuracy. The key idea is performing stereo-matching by dividing the rectified stereo pair into smaller overlapping tiles, which are processed hierarchically with iterative refinements.

HITNet follows the principles of traditional matching methods as three step process: compact feature representation, high resolution disparity initialization from features and efficient propagation to refine the estimates using support windows. The feature extraction module is a U-Net like architecture. During initialization, matches for all disparities are computed exhaustively. The index location of the best match is stored. During the propagation step, the input consists of tile hypotheses, and the output is refined tile hypotheses. This refinement is achieved by spatially propagating and fusing information by warping features from the feature extraction stage to predict highly accurate offsets to the input tiles. This approach allows for flexible, learned representations and provides good results.

3.3.3 RAFT

RAFT stereo is similar to RAFT for optical flow problems, in depth estimation it constructs a 4D cost volume from the correlation between pixels. It consists of three main components: (1) a feature and context encoder to extract feature vectors; (2) a correlation layer for a 4D correlation volume for all pairs of pixels; and (3) a recurrent GRU that updates a flow field. The network consists of blocks to down sample and produces feature maps at 1/4 or 1/8 of the input image resolution. A 4-level pyramid using correlation volumes is constructed through repeated average pooling of the last dimension. The cost volume is filtered through a series of 3D convolutions before being mapped to a point-wise depth estimate. The correlation, disparity, and context features are concatenated and injected into a hidden state, which is further used to predict the disparity update, and retrieved values are concatenated into a single feature map. Up sampling the obtained disparity to match the ground truth disparities is recommended, but it works by converting the input to low resolution. This is a very fast approach and yields convincing results.

4 Experimental Results and Analysis

4.1 Evaluation Parameters

Various error metrics have been used to compare the disparity maps obtained from different SGBM variants, the learning-based approach, and the optimization-based PatchMatch technique. These metrics help in making an informed decision about different approaches and their performance in different kinds of regions, such as flat/textured/texture-less regions.

- Mean Squared Error (MSE). Mean squared error (MSE) is a simple and intuitive measure to quantify the difference between the ground truth disparity map and the obtained disparity map and is represented as:

$$MSE = \frac{1}{N}\sum\nolimits_{i=1}^{N}(x_i - y_i)^2 \tag{7}$$

  In this equation, x_i and y_i are the images being compared with N number of pixels. The errors are squared to amplify error values.
- Bad Matching Pixels (BMP). It refers to the total count of pixels in an image that has a corresponding pixel that is significantly different in terms of color or texture. Bad matching pixels can occur for several reasons, including occlusions, specular reflections, texture-less regions, and errors in the stereo-matching algorithm. This metric is used to compare different traditional and learning approaches. Bad-matched pixel is defined as:

$$B(i,j) = \{-1, \mathit{if}\,|D(i,j) - G(i,j)| > t; 0, \mathit{otherwise} \tag{8}$$

  Here, $D(i,j)$ is the obtained disparity map, and $G(i,j)$ refers to the ground truth.
- Structural Similarity Index (SSIM). The Structural Similarity Index (SSIM) is a widely used metric in computer vision for measuring the similarity between two images. structural similarity is defined as:

$$SSIM\,(x,y) = \left[I(x,y)^{\alpha}[C(x,y)^{\beta}[S(x,y)^{\gamma}]]\right] \tag{9}$$

  where x and y are the two images being compared and are the constants that control the relative importance of these three terms. The terms *I(x, y)*, *C(x, y)*, and *S(x, y)* are the local luminance, contrast, and structural similarities, respectively, which are computed over a small window of pixels. SSIM values range between 0 and 1, where a value of 1 indicates perfect similarity and a value of 0 indicates non-similarity.

5 Result and Analysis

In this section, the dense disparity estimation techniques for aerial images are compared quantitatively and qualitatively. The standard error metrics such as mean square error (MSE), structure similarity index measure (SSIM), and bad matched pixels (BMP) have been used to benchmark the performance of different algorithms used in this work. The overall pipeline for code development and testing was carried out on a system with 4 GB RAM and an i3 processor. The SGBM algorithm was implemented using the source code of OpenCV, and the same was modified to compare the performance with different cost functions while keeping the remaining pipeline exactly the same. The source codes for the CNN-based architectures HITNet, CRE, and RAFT have been taken from their respective repositories.

The two datasets that are considered here for experimentation are the WHU dataset and the Mid-Air dataset. The WHU stereo disparity dataset consists of 1700 images divided into two subsets: the training and the testing set. These images were acquired using the GF-7 satellite covering various landscapes, including urban areas, rural areas, forests, and mountains. The Mid-Air dataset is an aerial stereo dataset that provides a large number of stereo images captured in a synthetic environment wherein the images are captured from a drone in different settings. It contains high-resolution stereo pairs with a large baseline and wide field of view. It also provides accurate ground truth disparity maps, which can be used for evaluating the accuracy of the algorithms. Both the datasets provide left image, right image and the corresponding disparity map. Including such datasets in evaluation not only helps in comparative analysis but also provides a benchmark to improve and enhance the quality of the algorithm. The dataset contains images with different trajectories and weather conditions, as seen visually. The images in the WHU dataset are divided into three categories, buildings, trees, and mixed regions, whereas for the Mid-Air dataset, they are divided into trees, flat regions, and rocky regions. A small proportion of images from each category is selected in a random manner to cater to all the terrains.

5.1 Quantitative Analysis

The objective behind this analysis is to find a technique that yields promising results for different kinds of aerial images with a balanced trade-off between time, complexity, accuracy, and overall quality of the depth map. The disparity range of aerial images is relatively large as compared to the non-aerial images due to the greater variation in the kinds of objects seen by the top-down camera, such as buildings, trees, flat surfaces, etc. It has been found during experimentation that the disparity values obtained using different techniques had different disparity ranges. They have, therefore, been compared in non-normalized and normalized modes so that their accuracy can be compared. It can also be argued that a comparison of the raw results would provide a better analysis of various techniques and are therefore included in some of the metrics, as will be discussed in the following subsections.

MSE. Tables 1 and 2 compare the MSE value for disparity obtained from different techniques in two different datasets, that is, the WHU and Mid-Air datasets considered here. In the case of the WHU and Mid-Air datasets, the MSE value between the ground truth and non-normalized disparity value is found to be very high as compared to the normalized cases. Normalization of the disparity value is carried out in the range of 0–75, which is chosen as the median of minimum and maximum disparity values in the ground truth images for the WHU dataset. In the case of the Mid-Air data set, the upper range value was 81 but has been chosen as 75 here for the sake of simplicity. There are some interesting inferences from this table which are listed as follows:

1. MSE values for different variants of SGBM are similar for WHU and Mid-Air datasets and are lower among all for the normalized cases as compared to nonnormalized cases.
2. MSE values for all three learning-based techniques are relatively better than SGBM variants in texture-less and synthetic images.
3. PatchMatch as a technique performs better across all the different kinds of images and has better MSE than all the other methods. This can be attributed to the solution of disparity value obtained by solving an optimization function.
4. In the case of mixed regions and region with buildings, SGBM-ADC provides better results than SGBM-BT, SGBM-SAD, and other learning-based approaches in the case of non-normalization.

Overall, it can be summarized that for flat regions, the results of SGBM are not very good since these regions consist of similar texture, color information, and no distinct edges or occlusions. SGBM variants perform equally well for images with good texture, such as building and mixed regions. The SGBM-based approach works

Table 1 Comparison of MSE for WHU dataset using different DE techniques

Disparity technique	WHU dataset non-normalized				WHU dataset normalized			
	Building	Flat	Mixed	Overall	Building	Flat	Mixed	Overall
SGBM-ADC	104.7	110.6	99.63	103.37	71.55	83.7	66.59	81.7
SGBM-SAD	109.2	106.5	104.79	103.95	66.54	80.8	67.16	70.5
SGBM-BT	105.9	117.8	105.98	106.21	72.78	83.3	65.87	71.34
CRE	113.6	107.0	116.8	108.86	102.25	99.0	103.2	95.56
HITNet	105.3	100.6	108.21	113.19	87.85	91.9	101.2	108.5
RAFT	109.3	106.2	102.62	104.14	104.94	99.8	100.9	101.9
PM	105.1	57.54	99.66	97.23	59.24	69.1	53.91	59.86

Table 2 Comparison of MSE for mid-air dataset using different DE techniques

Disparity technique	Mid-air dataset non-normalized				Mid-air dataset normalized			
	Trees	Flat	Rocky	Overall	Trees	Flat	Rocky	Overall
SGBM-ADC	126.2	130.3	138.9	121.6	108.1	108.8	95.45	107.9
SGBM-SAD	122.1	130.2	136.4	122.8	103	118.6	85.65	105.9
SGBM-BT	121.0	131.7	139.7	123.6	106.9	93.45	89.15	106.9
CRE	90.86	79.6	81.12	89.17	85.05	86.98	74.12	79.7
HITNet	86.33	88.6	84.27	90.77	88.81	85.12	80.46	86.68
RAFT	110.7	115.9	119.7	112.3	112.3	110.0	85.73	107.2
PM	89.34	87.3	84.19	87.27	86.68	84.46	79.79	84.09

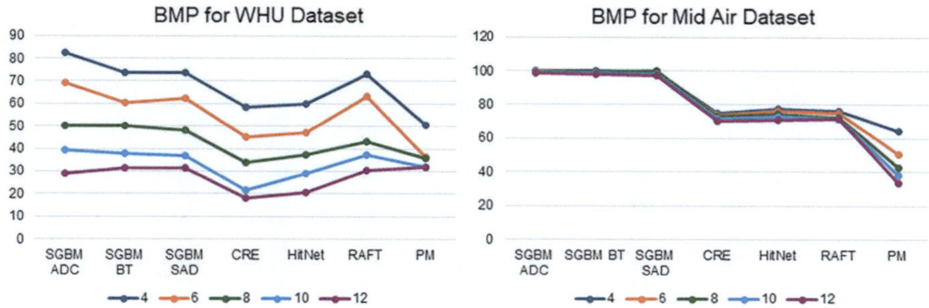

Fig. 2 BMP values for WHU and mid-air dataset

well for real-world images in the WHU dataset, while the learning-based approach works better for synthetic images in the Mid-Air dataset. However, MSE has certain limitations; it is sensitive to small variations and does not consider the perceptual quality of the constructed disparity map.

Bad Matched Pixels (BMP). It can be seen in Fig. 2 that the BMP values for both the WHU dataset and the Mid-Air dataset are found to decrease as the threshold value increases. In the case of the WHU dataset, the traditional SGBM-based approach is found to have a relatively higher error percentage as compared to the learning-based approaches. The performance of RAFT is comparable to other approaches, and the performance of SGBM—ADC is better than that of other SGBM variants. The performance of the patch match is better than that of all other techniques using an optimization approach. In the case of the Mid-Air dataset the performance of SGBM is similar.

Structural similarity index. Figure 3 compares the performance of different disparity estimation techniques in both the WHU and Mid-Air datasets. The SSIM value for all the schemes is higher in the case of the WHU dataset as compared to Mid—Air dataset. The traditional SGBM-based approach performs convincingly as compared to the learning-based approach in the case of the WHU dataset, while the same is not the case in the Mid-Air dataset. In learning-based techniques, these images are processed in different scales and have an output that is smoothed out such that there are no sudden changes in disparity levels. The SGBM-based techniques can preserve the minute details to match left and right images while reducing the overall structural information.

5.2 *Qualitative Results*

Analysis on WHU dataset. The images from three different categories have been tested using different disparity estimation schemes and shown in Fig. 4. Figure 4a–c represents the original left image as provided in the WHU dataset, Row 2 (second row) corresponds to the ground truth disparity map provided by the publishers.

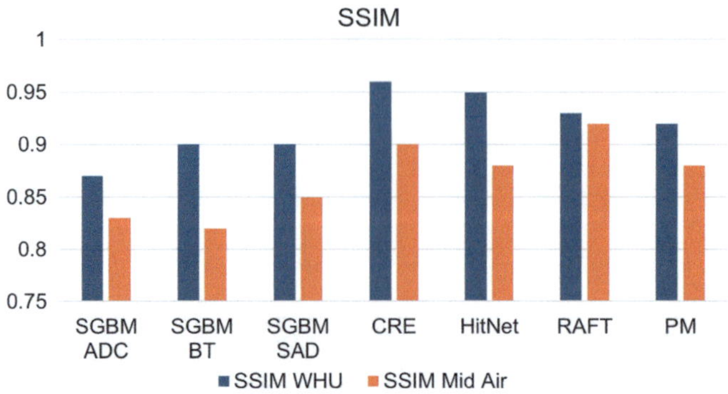

Fig. 3 SSIM comparison for WHU and mid-air datasets

Row 3–Row 5 presents disparity estimation results using different SGBM variants, Row 6–Row 8 benchmarks the performance of learning-based disparity estimation approaches while Row 9 depicts the performance of optimization-based Patch-Match approach. The disparity images shown here are in gray scale and normalized on a common scale to maintain uniformity. In Fig. 4d–f, the original left images from the Mid-Air dataset have been represented. In subsequent rows, the disparity output of different techniques can be observed.

The number of disparity levels has been configured to 96 for SGBM, which implies that for every pixel (x, y) in the left image, the algorithm calculates the cost for pixels ranging from (x − 96, y) to (x, y) in the right image. The number of disparities is a crucial parameter whose value, if set too low, can lead to a lot of noise and incorrect matches, especially in regions where the actual disparity is larger than the number of disparities set.

It can be seen from Fig. 4 that the SGBM does not yield as good a result as much as the deep learning frameworks, especially for images that are texture-less or contain too many discontinuities. The primary reason for this can be attributed to limited window size and local information available in window-based matching, unlike the learning-based approach that uses global information. In the case of images with well-defined structures, that is, buildings, represented as the leftmost image column, the performance of SGBM is relatively better as compared to other images, which have low texture areas. The zoomed-in portion of the building region is shown separately in Fig. 5 to compare the performance of different schemes. As seen in the zoomed-in region, for ADC, SGBM-BT, and SGBM-SAD, the edge information is not preserved to the same extent as that in SGBM—ADC as the latter is more robust over a window than the pixel-based intensity difference. The black portions beside the edges can be attributed to the mismatches. As compared to the SGBM-based approach, the learning-based methods show good performance, with RAFT preserving edge information of buildings better than HITNet, even for occluded cases. CRE provides distinct information about the structure and perception of depth is even better while the edges are smoothed out as compared to RAFT. Patch Match

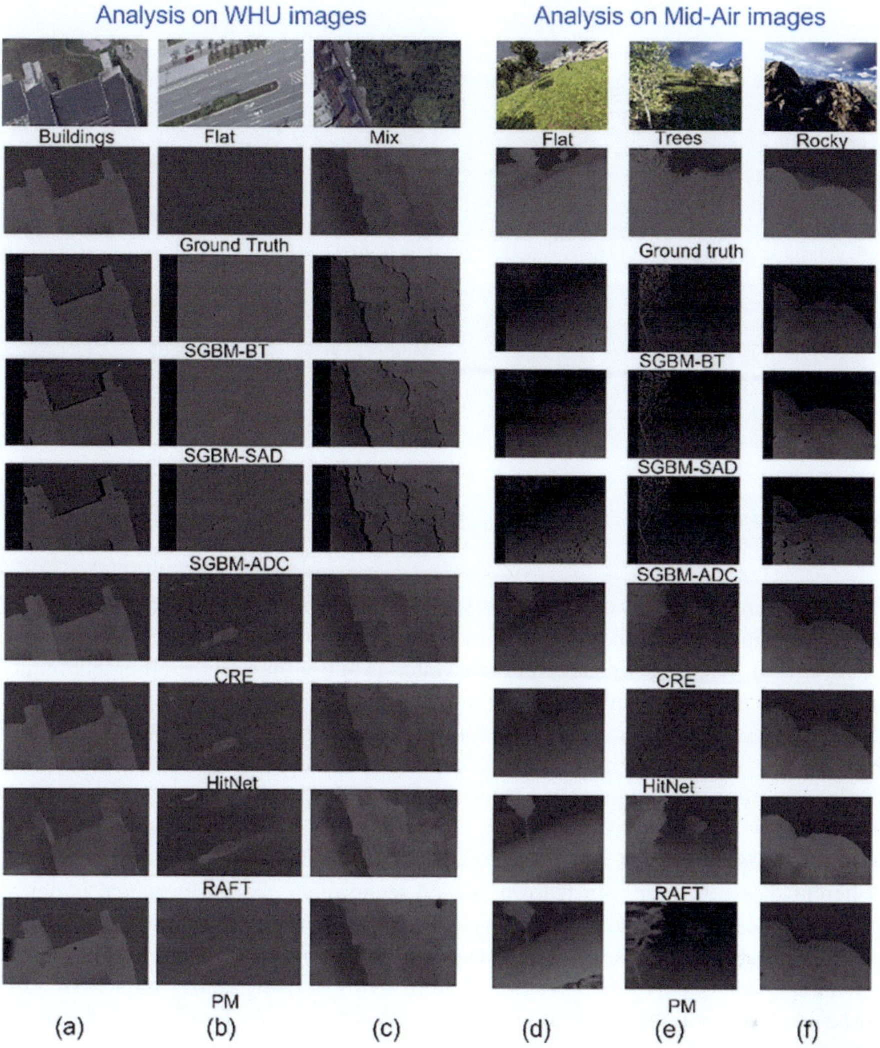

Fig. 4 Qualitative comparison of disparity estimated through different techniques on different images from WHU and mid-air dataset

produces visually good results except for corners where it is not able to handle edges and occlusion very well, as can be seen through a black blob in the Figure for the building case.

In Fig. 5a, the disparity maps for flat regions show that learning-based methods can detect edges and handle occlusions efficiently for wide texture-less regions as they are trained on several representations rather than pixel intensity differences. In this image, since most of the pixels have similar color information, it is difficult for the SGBM algorithm to find a reliable match, leading to blurring near occlusion

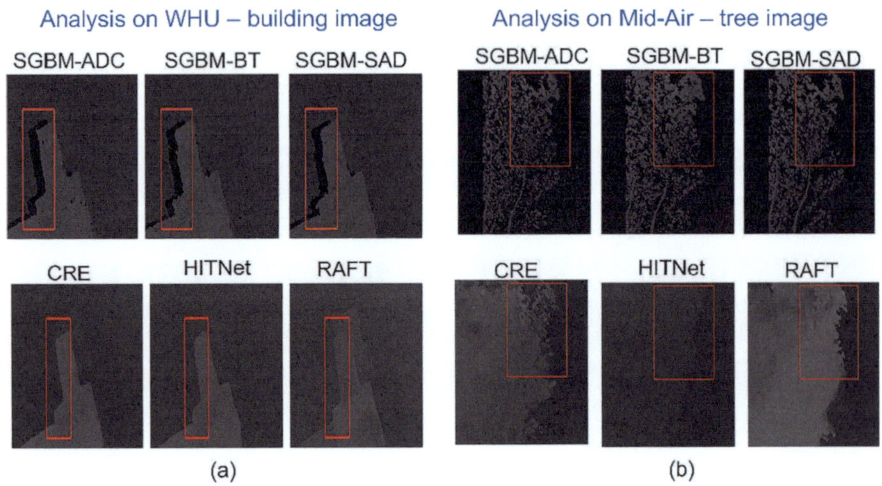

Fig. 5 Detailed analysis of border regions and flat areas for **a** WHU image **b** mid-air image

and discontinuities. This can be improved for SGBM schemes by a combination of pre-processing, post-processing, parameter tuning, and possibly using another type of algorithm. The performance of Patch Match is better than all the other techniques owing to its globally optimized result and consistency for all kinds of images.

In the case of images containing a lot of information, such as buildings, trees, etc., SGBM–ADC performs relatively better as the image contains different textures, distinct features, and edges. The performance of SGBM—BT and SGBM—SAD is not as good as SGBM—ADC. Also, during discontinuity or edges, SGBM uses a smoothness constraint that penalizes high-depth variations, producing visually coherent disparity maps. In learning-based techniques, the performance is not as good as expected, and the edges of the building cannot be appropriately deciphered from the results. In the case of CRE, it uses iterative refinement that allows it to process fine details and depth in dense regions, but in terms of using contextual information and global constraints, it is not visually as good as others. RAFT uses Recurrent Neural networks to refine disparity while using information from neighboring pixels as well and has relatively better performance than other learning approaches.

Analysis on Mid Air Dataset. This helps in understanding the performance of different techniques in different kinds of terrains and has been accordingly represented in Fig. 4d–f. The first row represents the original left image as captured by the UAV, and the remaining images are arranged similarly to the WHU dataset example. The disparity estimation result for images with trees can be found to be better for SGBM variants as compared to the learning-based approach. In the case of SGBM variants, the leaves in the tree and its leaves are visually distinguishable, which is not the case for the learning-based approach. In this case, even the ground truth does not provide detailed structure information compared to the original image. SGBM-BT yields a smooth output such that the background information is merged with the foreground information, which is not the case with SGBM-ADC as it is effective in handling texture-less regions in this case and can be found to provide better information about the ground as compared to the SGBM-BT and SGBM-SAD results. In the case of learning-based methods, CRE-stereo performs better than other learning-based schemes but not as good as the SGBM variants and can be inferred from the cropped leaf region. RAFT produces a disparity map which is smoothed out such that the leaves are not distinctly visible and not as sharp as compared to CRE. HITNet provides similar results as RAFT and gives a rough indication about trees/ vegetation with edges smoothed out. Patch-Match can provide clean and visually acceptable information about the scene due to its iterative plane refinement nature. It can distinguish between the flat regions and trees but the edges and texture information about the tree is lost in the process.

For the rocky areas in Fig. 4e, SGBM variants can detect edges with distinct boundaries and handles depth complexities well except in some equi-depth regions. For these kinds of images, the distinct edges encompassing the boundary region are seen properly, while the inner silhouette is not as distinct as the original image. In the case of SGBM—ADC, there are certain black spot regions, indicating no-matching availability on the corresponding image, which is otherwise smoothened out in the case of SGBM-BT and SGBM—SAD using a median and speckle filter. In the case of a learning-based method such as HitNet, it does not give accurate information about edges and is unable to separate the foreground and background regions accurately due to image complexity. CRE and RAFT produce relatively better depth maps as compared to HITNet, as they can distinguish between depth variations in different areas and maintain the edge information. In the case of PatchMatch, the variations in the depth map are better than those of SGBM variants and other deep learning methods.

The flat region in the Mid-Air dataset shown in Fig. 4f is a set of images that consists of large patches of areas having similar textures and constant depth variation. The performance of SGBM—ADC is found to be better than SGBM—BT and SGBM—SAD, depicting clear boundaries, especially for nearby objects, and fading away as the object gets farther away from the image viewpoint. In learning-based techniques, degraded performance is seen for nearby and far-away objects such that the ground merges with the sky when they are far away and cannot be distinguished in many cases. In the case of CRE and HITNet, the output is blurred near the edges and values. In some cases, the sky and flat ground have been assigned similar disparity, resulting in blurring and smoothing near surface boundaries, which is not the case with RAFT. PatchMatch can show distinct features in texture-less regions and yields better results as compared to others.

Overall, it can be inferred from the visual analysis that SGBM variants produce less accurate results near depth discontinuities and occluded regions compared to the other areas in the image. On the other hand, SGBM—ADC generates relatively accurate disparity values near the depth edges compared to the other regions in the image and tends to preserve the occluded regions. On analyzing the results of different learning-based approaches, it can be seen that these algorithms are able to yield a good estimate of image semantics but still need more tuning and enhancements to yield results that are comparable to optimization-based techniques. CRE and RAFT are more suitable with low-textured areas or repetitive texture surfaces like wood, grass, etc., and can handle the occlusions well. The disparity maps generated by HITNet, at times, lack some necessary details, which can be crucial for some applications. CRE gradually increases resolution during iterative updates, leading to a loss of context during information propagation, leading to image blurring at different instances. It can be inferred from visual analysis that the performance of optimization techniques, that is, PatchMatch used here is better in most of the cases but at a very high computational cost. It is, therefore, necessary that faster computational algorithms be developed that take cues from computer vision geometry and do not rely on learning-based architectures alone.

5.3 Computation Time Analysis

The computation time of any algorithm is one of the major factors governing usage in real-time applications to compare the different methods experimented here. The time required to run different techniques on the two datasets for a single image pair is observed and plotted in Fig. 6. In the case of SGBM variants, it was found that changing cost functions led to minor changes in computational time. The optimization-based approach PatchMatch takes a relatively much longer time than the traditional SGBM-based or learning-based approaches. It is for this very reason that despite yielding good results, they are not practically realizable for real-time systems. The time taken for the Mid-Air dataset is higher as the image resolution is higher than the WHU dataset. These implementations are not optimized for parallel processing or run-on optimized hardware resources, which could have led to less computation time in each of the cases.

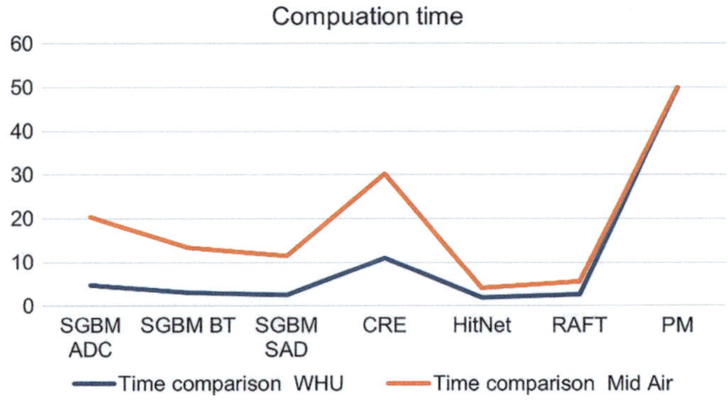

Fig. 6 Computation time analysis of different disparity estimation techniques

6 Conclusion

In this work various metrics to compare disparity maps obtained from different traditional and learning based approaches have been used. It was found that the image texture and features play a very important role in the estimated disparity. The block matching method primarily utilizes low-level image features to identify corresponding pixels in the left and right images. As a result, the disparity maps generated by this method are often noisy and lead to decreased performance. However, despite this limitation, the method still produces disparity maps that preserve geometric accuracy and the performance is identical with different cost functions.

In case of learning-based approaches, they are able to carry out the semantic segmentation appropriately. However, their MSE values differ widely from the ground truth image as they do not have any common reference. The dense depth map in case of SGBM based techniques are obtained by calculating the shift in disparity between left and right image, while in case of learning-based approach, these values are based on a pre-trained model. Therefore, when the same approach is applied on a new kind of dataset, normalization of the result within a specified disparity level is very necessary. The learning-based approach gives a very good approximation of the relative disparity, and if a hybrid approach is created to fuse the computational geometry like SGBM, it might yield better results than either of the two approaches.

Among the learning-based methods, CRE Stereo was found to be a robust algorithm that performed well in almost all the cases. It is memory efficient as it uses a local search window instead of a full-cost volume. HITNet avoids full-cost volume computation and adapts a coarse-to-fine propagation approach. Raft Stereo is fast, accurate, and provides good results but blurs textures and occlusions. In the context of aerial images, the choice of technique for disparity depends on various factors like quality of images, scene complexity, and color or gradient information. It should be noted that different methods can be tuned to specific applications or a particular dataset and for different regions. However, more efforts are required to develop algorithms that are both geometrically consistent and fast.

Conflict of Interest The authors declare that they have no conflict of interest.

Funding The funding for this project has been received from DRDO—Aeronautical Research and Development board.

Availability of data and materials Data sharing not applicable to this article as no datasets were generated. The analysis of results have been provided in the manuscript.

Code availability: Can be provided on request basis.

Ethics approval Not applicable.

References

1. Baidya R, Jeong H (2024) Simulation and real-life implementation of UAV autonomous landing system based on object recognition and tracking for safe landing in uncertain environments. Front Robot AI 11:1450266
2. Tovanche-Picon H, González-Trejo J, Flores-Abad Á, García-Terán MÁ, Mercado-Ravell D (2024) Real-time safe validation of autonomous landing in populated areas: from virtual environments to robot-in-the-loop. Virt Real 28(1)
3. Narayan I, Battish N, Kaur D, Gupta A, Poddar S (2024) Landing site selection for UAV in unknown environment using surface inclination. SPIE Future Sens Technol 13083:140–149. https://www.spiedigitallibrary.org/conference-proceedings-of-spie/13083/130830W/Landing-site-selection-for-UAV-in-unknown-environment-using-surface/10.1117/12.3023755.short
4. Dhrafani D, Liu Y, Jong A, Shin U, He Y, Harp T, Hu Y, Oh J, Scherer S (2024) FIReStereo: forest InfraRed stereo dataset for UAS depth perception in visually degraded environments. arXiv:2409.07715, https://doi.org/10.48550/arXiv.2409.07715
5. Li S, He S, Jiang S, Jiang W, Zhang L (2023) WHU-stereo: a challenging benchmark for stereo matching of high-resolution satellite images. IEEE Trans Geosci Remote Sens 61:1–14
6. Kordelas GA, Alexiadis DS, Daras P, Izquierdo E (2015) Enhanced disparity estimation in stereo images. Image Vis Comput 35:31–49
7. Zeng H, Ren J, Qin Y (2023) Research based on improved AD-census stereo matching algorithm. In: Proceedings of the 2023 3rd International Conference on Bioinformatics and Intelligent Computing, pp 88–92. https://doi.org/10.1145/3592686.3592703
8. Liu B, Yu H, Long Y (2022) Local similarity pattern and cost self-reassembling for deep stereo matching networks. In: Proceedings of the AAAI conference on artificial intelligence, vol 36(2), pp 1647–1655. https://ojs.aaai.org/index.php/AAAI/article/view/20056
9. Sun J, Zheng N-N, Shum H-Y (2003) Stereo matching using belief propagation. IEEE Trans Pattern Anal Mach Intell 25(7):787–800
10. Yang Q, Wang L, Yang R, Stewénius H, Nistér D (2008) Stereo matching with color-weighted correlation, hierarchical belief propagation, and occlusion handling. IEEE Trans Pattern Anal Mach Intell 31(3):492–504
11. Bleyer M, Rhemann C, Rother C (2011) Patchmatch stereo-stereo matching with slanted support windows. BMVC 11:1–11. https://bmva-archive.org.uk/bmvc/2011/proceedings/paper14/paper14.pdf
12. Yang S, Lei X, Liu Z, Sui G (2021) An efficient local stereo matching method based on an adaptive exponentially weighted moving average filter in SLIC space. IET Image Proc 15(8):1722–1732. https://doi.org/10.1049/ipr2.12140
13. Žbontar J, LeCun Y (2016) Stereo matching by training a convolutional neural network to compare image patches. J Mach Learn Res 17(65):1–32
14. Jia Q, Wan X, Hei B, Li S (2018) DispNet based stereo matching for planetary scene depth estimation using remote sensing images. In: 2018 10th IAPR workshop on pattern recognition in remote sensing (PRRS), pp 1–5. https://ieeexplore.ieee.org/abstract/document/8486195/
15. Guo X, Yang K, Yang W, Wang X, Li H (2019) Group-wise correlation stereo network. In: Proceedings of the IEEE/CVF conference on computer vision and pattern recognition, pp 3273–3282. http://openaccess.thecvf.com/content_CVPR_2019/html/Guo_Group-Wise_Correlation_Stereo_Network_CVPR_2019_paper.html
16. Tankovich V, Häne C, Zhang Y, Kowdle A, Fanello S, Bouaziz S (2020) HITNet: hierarchical iterative tile refinement network for real-time stereo matching (version 5). https://doi.org/10.48550/ARXIV.2007.12140
17. Xu H, Zhang J (2020) AANET: adaptive aggregation network for efficient stereo matching. In: Proceedings of the IEEE/CVF conference on computer vision and pattern recognition, pp 1959–1968. http://openaccess.thecvf.com/content_CVPR_2020/html/Xu_AANet_Adaptive_Aggregation_Network_for_Efficient_Stereo_Matching_CVPR_2020_paper.html

18. Kendall A, Martirosyan H, Dasgupta S, Henry P, Kennedy R, Bachrach A, Bry A (2017) End-to-end learning of geometry and context for deep stereo regression. In: Proceedings of the IEEE international conference on computer vision, pp 66–75. http://openaccess.thecvf.com/content_iccv_2017/html/Kendall_End-To-End_Learning_of_ICCV_2017_paper.html
19. Chang J-R, Chen Y-S (2018) Pyramid stereo matching network. In: Proceedings of the IEEE conference on computer vision and pattern recognition, pp 5410–5418. http://openaccess.thecvf.com/content_cvpr_2018/html/Chang_Pyramid_Stereo_Matching_CVPR_2018_paper.html
20. Lipson L, Teed Z, Deng J (2021) RAFT-Stereo: multilevel recurrent field transforms for stereo matching. In: 2021 international conference on 3D vision (3DV), pp 218–227. https://doi.org/10.1109/3DV53792.2021.00032
21. Wu Z, Wu X, Zhang X, Wang S, Ju L (2019) Semantic stereo matching with pyramid cost volumes. In: Proceedings of the IEEE/CVF international conference on computer vision, pp 7484–7493. http://openaccess.thecvf.com/content_ICCV_2019/html/Wu_Semantic_Stereo_Matching_With_Pyramid_Cost_Volumes_ICCV_2019_paper.html
22. Haeusler R, Klette R (2012) Analysis of KITTI data for stereo analysis with stereo confidence measures. In: Fusiello A, Murino V, Cucchiara E (eds) Computer vision – ECCV 2012. Workshops and demonstrations, vol 7584. Springer Berlin Heidelberg, pp 158–167. https://doi.org/10.1007/978-3-642-33868-7_16
23. Hirschmuller H (2007) Stereo processing by semiglobal matching and mutual information. IEEE Trans Pattern Anal Mach Intell 30(2):328–341
24. Heo YS, Lee KM, Lee SU (2010) Robust stereo matching using adaptive normalized cross-correlation. IEEE Trans Pattern Anal Mach Intell 33(4):807–822
25. Huang H, Yan X, Zheng Y, He J, Xu L, Qin D (2024) Multi-view stereo algorithms based on deep learning: a survey. Multimed Tools Appl 84(6):2877. https://doi.org/10.1007/s11042-024-20464-9
26. Laga H, Jospin LV, Boussaid F, Bennamoun M (2020) A survey on deep learning techniques for stereo-based depth estimation. IEEE Trans Pattern Anal Mach Intell 44(4):1738–1764
27. Li L, Yu X, Zhang S, Zhao X, Zhang L (2017) 3D cost aggregation with multiple minimum spanning trees for stereo matching. Appl Opt 56(12):3411–3420
28. Scharstein D, Szeliski R (2002) A taxonomy and evaluation of dense two-frame stereo correspondence algorithms. Int J Comput Vis 47(1/3):7–42. https://doi.org/10.1023/A:1014573219977
29. Shi B, Shi S, Wu J, Chen M (2019) A new basic correlation measurement for stereo matching. OSF Preprints, 29 May 2019. https://files.osf.io/v1/resources/jqux2/providers/osfstorage/5cee659e2a50c40019814be9?action=download&version=1&direct&format=pdf
30. Wang Y, Gu M, Zhu Y, Chen G, Xu Z, Guo Y (2022) Improvement of AD-census algorithm based on stereo vision. Sensors 22(18):6933
31. Zhang X, Cao X, Yu A, Yu W, Li Z, Quan Y (2023) UAVStereo: a multiple resolution dataset for stereo matching in UAV scenarios. IEEE J Sel Top Appl Earth Observ Remote Sens 16:2942–2953

Machine Learning Models-Based Prediction in Cardiovascular Diseases: A Cavernous Analysis

Anu Singha, Mehul Bhatia, Prajayshee Chauhan, and Sakshi Jaiswal

Abstract Cardiovascular (CVD) disease continues to pose a key global health challenge, emphasizing the need for accurate risk prediction and preventive measures. In recent decades, machine learning (ML) approaches have arisen as influential tools for analyzing complex medical datasets and enhancing CVD risk assessment. This book chapter presents a comprehensive review of recent advancements in ML-based CVD prediction, covering various ML algorithms, datasets, feature selection techniques, performance evaluation metrics, and associated challenges. The healthcare industry generates vast amounts of medical data, necessitating ML-driven decision-making for effective heart disease prediction. Recent research has explored the integration of multiple ML techniques to develop hybrid predictive models for improved accuracy. The proposed study employs data pre-processing techniques such as noise removal, handling missing values, and attribute classification to enhance classification and decision-making at different stages. The performance of the predictive model is assessed using classification metrics such as sensitivity, accuracy, and specificity. This chapter introduces a CVD prediction classification model designed to determine the likelihood of heart disease and raise awareness regarding early diagnosis. The proposed approach compares the predictive accuracy of decision tree, random forest, gradient boosting, and logistic regression by applying rule-based methodologies to regional datasets, ultimately identifying the most accurate model for CVD prediction.

Keywords Prediction · Healthcare · Machine learning algorithms · Cardiovascular disease

A. Singha (✉) · M. Bhatia · P. Chauhan · S. Jaiswal
Dr. Vishwanath Karad MIT World Peace University, Pune, India
e-mail: anu.singha@mitwpu.edu.in

M. Bhatia
e-mail: mehul.bhatia@mitwpu.edu.in

P. Chauhan
e-mail: prajayshee.chauhan@mitwpu.edu.in

S. Jaiswal
e-mail: sakshi.jaiswal@mitwpu.edu.in

S. Chand Kumain et al. (eds.), *AI and ML Techniques in Image Processing and Object Detection*, https://doi.org/10.1007/978-981-96-7445-9_8

1 Introduction

CVD refers to a broad spectrum of circumstances affecting the blood vessels and heart, including stroke, coronary artery disease, and heart attack. Despite substantial advancements in medical treatment and public health initiatives, CVD remains the leading global cause of mortality. Early detection of patients at peak risk is essential for implementing effective protective measures and improving patient outcomes. Conventional risk assessment models, which rely on demographic and clinical factors, often lack the precision needed for personalized risk prediction. In contrast, machine learning (ML) techniques have demonstrated the ability to analyze large-scale datasets and incorporate diverse risk factors, enhancing predictive accuracy in CVD assessment.

According to reports by the World Health Organization (WHO), over 12 million demises arise annually due to cardiovascular diseases, making it one of the most devastating global health concerns. In India, the impact of CVD is particularly severe, posing significant health risks. Diagnosing these conditions is a complex process requiring high precision, yet the shortage of medical experts in certain regions places patients at an increased risk of misdiagnosis. Typically, cardiologists diagnose and treat heart diseases, but integrating ML techniques with medical information systems can enhance diagnostic accuracy and bridge the gap in healthcare accessibility.

This paper explores various ML techniques used for predicting cardiovascular disease risk, comparing their effectiveness in analyzing uncertainty levels based on patient attributes. The study utilizes medical datasets collected from global research efforts to evaluate the performance of different ML models. Machine learning, which enables systems to learn patterns from data without explicit programming, has become a powerful tool for identifying complex patterns in high-dimensional, diverse datasets, such as those related to heart diseases. By leveraging these advanced techniques, CVD prediction models can achieve greater precision, contributing to early diagnosis and improved patient care.

2 Related Work

This section presents a comprehensive literature review of recent studies utilizing ML methods for cardiovascular disease (CVD) prediction. It examines commonly used ML techniques, including decision trees, logistic regression, random forests, support vector machines, and neural networks, highlighting their advantages and limitations. Additionally, it explores feature selection techniques such as filter, wrapper, and embedded approaches, which enhance model performance and interpretability. The role of integrating diverse data sources, including genetic data, electronic health records, imaging, and wearable sensor data, is also discussed to illustrate improvements in predictive accuracy.

Weng et al. [1] evaluated four ML models on clinical data from over 300,000 UK households, revealing that neural networks were the most effective for CVD prediction in large datasets. Saboor et al. [2] applied nine ML classifiers (e.g., SVM, Random Forest, Decision Tree, XGBoost) on a cardiovascular disease dataset, achieving 96.72% accuracy with SVM after data normalization and hyperparameter tuning. Similarly, Zaman et al. [3] proposed an IoT-ML model that analyzes heart rate, ECG signals, and cholesterol levels to assess cardiovascular health conditions. A meta-analysis by Krittanawong et al. [4] reviewed ML models for predicting coronary artery disease, heart failure, stroke, and arrhythmias using MEDLINE, Embase, and Scopus databases. Their findings emphasized that SVM and boosting algorithms demonstrated strong predictive performance, but significant variations were observed across different studies due to parameter inconsistencies. Bharti et al. [5] employed deep learning (DL) models on the Supervised Learning Archive Coronary Heart Disease dataset, achieving an average accuracy of 94.2% using 14 essential clinical features. Ahmed et al. [6] compared Logistic Regression, Decision Tree, Random Forest, and SVM for stroke prediction, with Random Forest achieving the highest accuracy (90%) after hyperparameter tuning and cross-validation. Similarly, Biswas et al. [7] tackled class imbalance in stroke diagnosis by applying Random Over Sampling (ROS) and analyzing eleven ML classifiers. After balancing the dataset, four classifiers exceeded 96% accuracy, demonstrating the effectiveness of oversampling techniques. Recent advancements have also explored cloud-based and computationally efficient models. Maini et al. [8] proposed a cloud-based decision support system for affordable CVD diagnosis using ML, while Enriko et al. [9] evaluated a K-Nearest Neighbors (KNN) model, achieving 81.85% accuracy but noting a decline in performance with an increased number of parameters. Anitha and Sridevi [10] utilized learning vector quantization algorithms on UCI's heart disease dataset, selecting 14 out of 76 features for improved prediction, achieving 85.55% accuracy.

This review highlights the increasing adoption of ML techniques for CVD prediction, demonstrating high accuracy and reliability in clinical settings. However, model interpretability, parameter optimization, and dataset variability remain key challenges, requiring further research to develop scalable and explainable AI-based healthcare solutions.

3 Dataset and Description

Data Source and Analysis: Healthcare databases have accumulated vast amounts of patient records, providing critical insights into various medical conditions, including heart disease. The term heart disease encompasses multiple conditions that adversely affect the human heart, with cardiovascular disease (CVD) being among the most severe. CVD refers to disorders impacting the heart and blood vessels, which disrupt blood circulation and pumping functions. For this study, records were sourced from the Cleveland, Switzerland, Hungarian, and Long Beach VA heart disease databases, available in the UCI Machine Learning Repository. These datasets are used to identify patterns associated with heart disease. The data is separated into two subsets: training and testing datasets. A total of 920 records with 76 medical attributes were collected, with 14 key attributes selected for analysis, as listed in Table 1.

Data Preprocessing: The data preprocessing stage plays a vital role in preparing the dataset for investigation. This involves:

- Data cleaning to eliminate inconsistencies.
- Data integration to merge relevant information.
- Handling missing values by either filling them with estimated values or removing incomplete records.
- Eliminating redundant data to prevent incorrect predictions.

Since missing or redundant data can lead to faulty predictions, preprocessing ensures that the dataset remains accurate and reliable for analysis.

Operating Environment and Data Analysis: The analysis was performed using Python, which provides a robust statistical computation and graphical representation platform for data-driven decision-making. Figures 1 and 2 illustrate insights derived from the dataset:

- Figure 1 highlights the relationship between age and cardiovascular disease occurrence, showing that individuals aged 55–65 are at the highest risk.
- Figure 2 depicts the impact of blood pressure levels, indicating that CVD risk is more prevalent in individuals with blood pressure readings between 110 and 150.

Table 1 Various attributes used are listed [11]. CN—Continuous

S. No.	Observation	Description	Values
i	Age	Age (years)	CN
ii	Sex	Patient sex	M/F
iii	CP	Chest pain	4 types
iv	Restbps	Resting blood pressure	CN
v	Chol	Cholesterol (serum)	CN
vi	FBS	Fasting blood sugar	<, or >120 mg/dl
vii	RestECG	Resting electro cardio gram	5 values
viii	Thalach	Maximum heart rate achieved	CN
ix	Exang	Exercise induced angina	Y/N
x	Oldpeak	ST depression during exercise in relation to the duration of rest taken	CN
xi	Slope	Slope of peak exercise ST segment	Up/Down/Flat
xii	Ca	Indicates the number of major vessels highlighted by fluoroscopy	0–3
xiii	Thal	Defect type	Reversible/Normal/Fixed
xiv	Num (disorder)	Heart disease	Not present/Present in the 4 major types

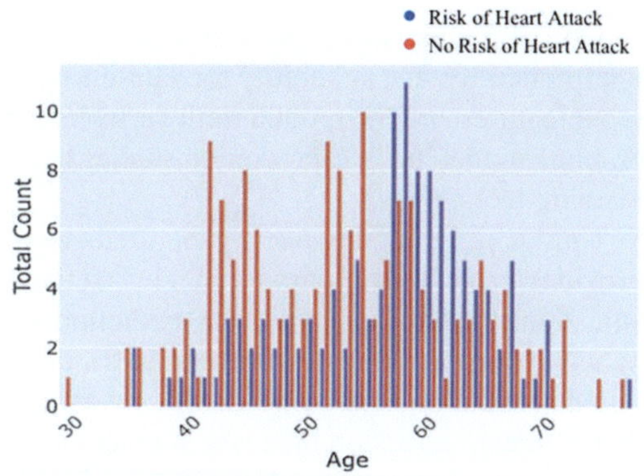

Fig. 1 Demonstrations the risk of heart attack based on age

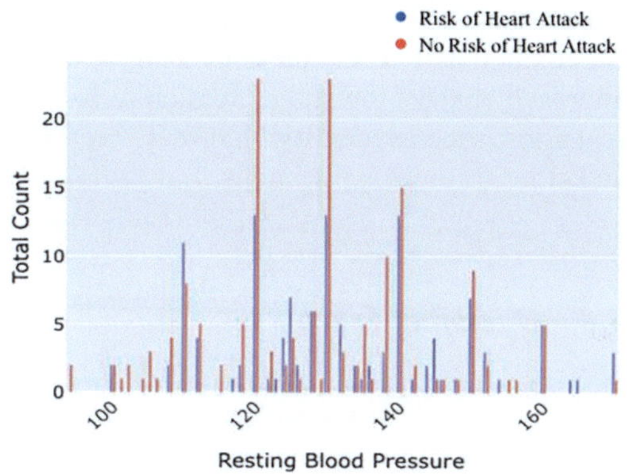

Fig. 2 Demonstrations the risk of heart attack based on resting blood pressure

Python's extensive libraries facilitate rapid data visualization and statistical analysis, enabling efficient development of predictive models for heart disease. While heart disease manifests in various forms, there are key risk factors that determine an individual's susceptibility. Monitoring these essential characteristics is crucial in assessing and predicting CVD risk [11].

4 Various Machine Learning Algorithms

4.1 Logistic Regression

Logistic regression [12] is a statistical procedure commonly used for binary classification tasks, such as determining the presence or absence of cardiovascular disease (CVD). It estimates the probability of a binary outcome based on one or more predictor variables, making it a widely used model in medical diagnostics.

One of the key advantages of logistic regression is its computational efficiency, interpretability, and suitability for datasets where a linear relationship exists between input features and the ground truth variable. Due to its simplicity, it often serves as a baseline model in CVD prediction studies before employing more complex machine learning techniques.

Unlike regression models that predict continuous values, logistic regression provides probabilistic outputs ranging between 1 and 0, rather than discrete classifications. For instance, in CVD prediction, the model estimates the likelihood of a patient having the disease (1: Presence) or not (0: Absence), instead of making absolute determinations. It achieves this by fitting an S-shaped logistic function rather than a linear regression line, which makes it well-suited for classification problems.

The sigmoid function, also recognized as the logistic function, is fundamental to logistic regression as it maps predicted values to probabilities. A threshold value is applied to classify outcomes: if the predicted probability exceeds the threshold, it is classified as 1 (CVD present); otherwise, it is classified as 0 (CVD absent). This approach enables logistic regression to be effectively used in various classification scenarios, such as identifying disease presence, assessing obesity based on weight, and other medical applications.

4.2 Decision Trees

Decision trees [13] are supervised learning non-parametric models that systematically divide the feature space into smaller subsets based on the values of input features. Each node in the tree signifies a decision criterion based on a particular feature, leading to an ultimate classification or prediction at the leaf nodes.

One of the key advantages of decision trees is their interpretability, as they provide a clear, step-by-step decision-making process. They can handle both categorical and numerical data, are resistant to outliers, and can manage missing values effectively. Additionally, decision trees can capture non-linear relationships between input features and the ground truth variable, making them well-suited for CVD prediction tasks.

Decision Tree Algorithm Workflow:

1. Initialize the Root Node: Start with the root node (S) containing the entire dataset.
2. Select the Best Attribute: Identify the most significant feature using an Attribute Selection Measure (ASM) such as Gini impurity or information gain.
3. Partition the Dataset: Divide S into subsets based on the possible values of the selected attribute.
4. Create Decision Nodes: Generate a decision node for the best attribute.
5. Recursive Tree Construction: Repeat the partitioning process for each subset until further classification is not possible. The final nodes are termed leaf nodes, representing the classification outcome.

By following this process, decision trees provide an intuitive and effective framework for diagnosing cardiovascular diseases, enabling automated decision-making based on patient attributes.

4.3 Random Forest

Random forests [14] are ensemble learning algorithms that association of several decision trees to enhance predictive accuracy and reduce the risk of overfitting. Each tree in the forest is trained on a random subset of the training data and features, and final predictions are obtained by aggregating the outputs of all individual trees.

One of the key strengths of random forests is their capability to grip high-dimensional datasets while maintaining robustness. Unlike individual decision trees, which are prone to overfitting, random forests provide better generalization performance by averaging multiple tree predictions. This makes them particularly effective for CVD prediction, where identifying feature importance and ensuring reliable model performance are crucial.

Random Forest Algorithm Workflow:

1. Random Data Selection: Choose K random data points from the training set.
2. Build Decision Trees: Construct decision trees based on the selected subsets of data.
3. Determine the Number of Trees: Set the number N of decision trees to be generated.
4. Repeat Steps 1 and 2: Continue building trees using different random subsets.
5. Prediction and Voting: For new input data, each decision tree makes a prediction. The final classification is determined by majority voting, where the category receiving the most votes is assigned to the new data point.

By leveraging multiple decision trees, random forests improve predictive accuracy, making them a robust and reliable method for diagnosing cardiovascular diseases.

4.4 Gradient Boosting Machines (GBM)

Gradient Boosting Machines (GBMs) [15] are ensemble learning methods that build a sequence of weak learners—typically decision trees—where each subsequent model aims to correct the errors of its predecessor. This iterative learning approach improves model performance by minimizing residual errors over multiple iterations.

Popular GBM frameworks such as XGBoost and LightGBM are highly effective in CVD risk prediction, as they efficiently process heterogeneous data, identify non-linear patterns, and achieve high predictive accuracy when fine-tuned with hyperparameter optimization. Their ability to adapt to misclassified instances makes them particularly well-suited for complex medical datasets.

Gradient Boosting Algorithm Workflow:

1. Initialize the Dataset—Start with a dataset containing multiple data points.
2. Assign Equal Weights—Each data point is initially given an equal weight.
3. Train the First Weak Learner—The model is trained using these weights as input.
4. Identify Misclassified Instances—Determine which data points were incorrectly predicted.
5. Adjust Weights—Increase the weights of misclassified data points so that subsequent models focus more on these cases.

By iteratively refining predictions, gradient boosting enhances model accuracy, making it a powerful approach for cardiovascular disease risk assessment.

5 Experiment Analysis and Result Discussion

In this section, we discussed experiment setup, assessment metrics, and comparative assessment. To train the machine learning methodologies, we have used a workstation server with 48 GB RAM, Intel Xeon CPU, CUDA 8.0, NVIDIA XP GPU implementation. In this chapter, 920 patients record samples are treated as train set and test along with 14 attributes.

5.1 Evaluation Parameter Discussion

Several essential metrics are available for evaluating machine learning models, and we have utilized five key measures.

ROC_AUC: The Receiver Operating Characteristic (ROC) curve is a probability-based evaluation metric, and the Area Under the Curve (AUC-ROC) quantifies the model's ability to distinguish between CVD and non-CVD cases. A higher AUC indicates better classification performance, as it represents the degree of separability between the two classes. The AUC is computed using the following formula:

$$AUC = \frac{\sum rank_i - \frac{M(M+1)}{2}}{MN} \tag{1}$$

$\sum rank_i$ denotes the total sum of the serial numbers assigned to cancer-positive samples. M and N correspond to the count of cancer-positive and cancer-negative samples, respectively.

The remining metrics formulas as follows:

$$Recall = \frac{TP}{TP + FN} \tag{2}$$

$$Precision = \frac{TP}{TP + FP} \tag{3}$$

$$F1\text{-}Score = 2 \times \frac{Precision \times Recall}{Precision + Recall} \tag{4}$$

$$Accuracy = \frac{TP + TN}{TP + TN + FP + FN} \tag{5}$$

where FP represents the count of false positive CVD cases, TN denotes the number of true negative CVD cases, TP refers to the number of true positive CVD cases, and FN indicates the count of false negative CVD cases, respectively.

5.2 Comparative Assessment

The comparative results section provides a performance assessment of various machine learning models in predicting cardiovascular disease (CVD) risk. This subsection evaluates and compares widely used models, including Decision Tree, Logistic Regression, Random Forest, and GBM. These models were tested using a heart disease dataset obtained from the UCI Machine Learning Repository. The classification performance of these models for CVD prediction is illustrated in Fig. 3.

In case of precision metric, the resultant metric reached up to 85% precision value which has achieved by Gradient Boosting Method. The performance of Random Forest also reached near to Gradient Boost. The poorest precision value is achieved by Decision Tree. In case of recall metric, the performance of Random Forest and Gradient Boosting approximately equal with approx. 83% recall values. And the lowliest recall value is attained by Logistic Regression. In case of F1-score, the outcomes of Random Forest and Gradient Boosting also approximately same with approx. 84% F1-scores. Here also, the poorest F1-score is achieved by Decision Tree. MCC provides a more informative and honest result when evaluating binary classifications compared to accuracy and F1-Score. The Gradient Boosting methods outperform all other methods.

An additional experimental evaluation was conducted to compare the performance of the listed methods. As shown in Fig. 4, the ROC curve plots the True Positive Rate against the False Positive Rate at different threshold values. When evaluating machine learning models using the ROC-AUC metric, the focus is on how effectively each algorithm distinguishes between the positive (CVD) and negative (non-CVD) classes. The AUC score quantifies the model's overall discriminative capability, with a higher value indicating better classification performance.

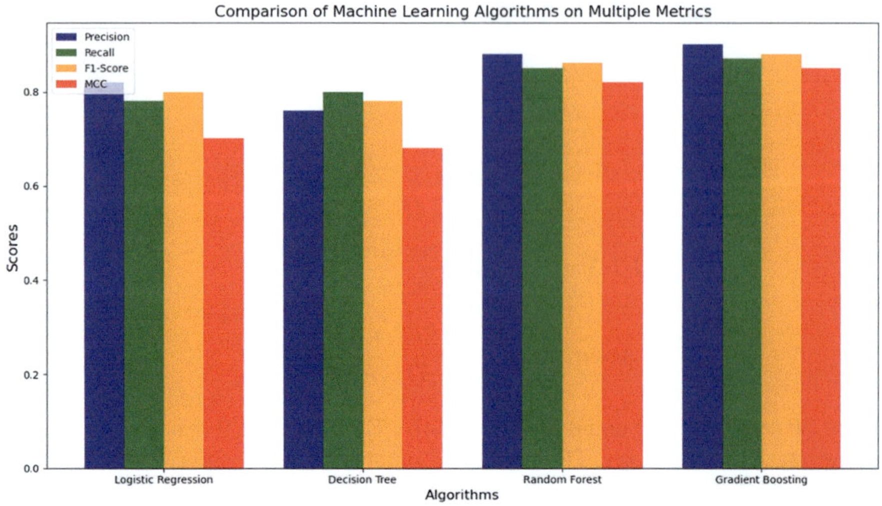

Fig. 3 Displays the risk of a heart attack based on resting blood pressure levels

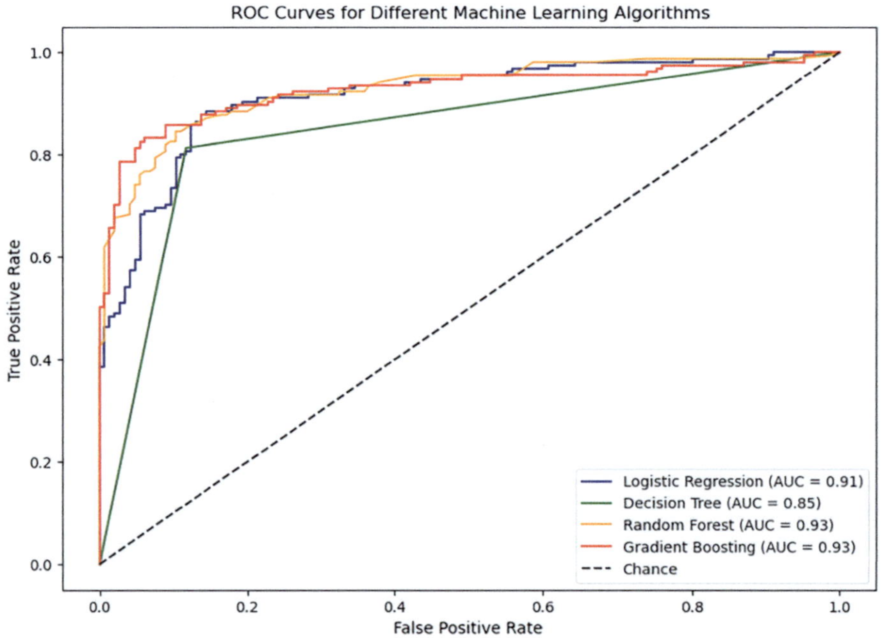

Fig. 4 Displays the risk of heart attack on the basis of their resting BP

Logistic Regression typically produces a smooth, sigmoid-shaped ROC curve, as it is a linear model that assumes a linear relationship between the input features and the log-odds of the outcome. This analysis helps in identifying the most effective model for CVD prediction based on classification performance and robustness. Logistic Regression has a high AUC (0.91), it suggests that the model is good at distinguishing between the two classes. However, since it's a linear model, its performance might be lower compared to more complex models if the true relationship between the features and the target is non-linear.

Decision Trees create a piecewise constant ROC curve because they partition the feature space into discrete regions. The ROC curve is more jagged due to the binary splits. The AUC for the Decision Tree is significantly lower (0.85) than that of other models, it may indicate that the model is overfitting. Decision Trees are prone to overfitting, especially on small datasets.

Random Forest, being an ensemble of decision trees, generally produces a smoother and more reliable ROC curve compared to a single decision tree. This model averages the predictions of multiple trees, reducing variance and improving generalization. The Random Forest has a high AUC (0.93), it indicates that the model is effectively capturing the complexities of the data. Typically, Random Forests outperform single Decision Trees because they mitigate overfitting and provide a more robust prediction.

GBM builds an ensemble of trees sequentially, where each tree attempts to correct the errors of the previous one. This often leads to a highly optimized and smooth ROC curve. GBM often achieves the highest AUC (0.93) among these models because it can capture complex patterns in the data. Since the Gradient Boosting has the highest AUC, it suggests that it is the best model for distinguishing between the classes in your dataset.

As a summary, the model with the highest AUC is typically considered the best at classifying the data, meaning it has the highest ability to distinguish between the positive and negative classes. Gradient Boosting often achieves the highest AUC due to its sequential, error-correcting approach. Random Forest is also a strong performer due to its ensemble nature, reducing overfitting. Models with lower AUC scores may be underfitting the data (failing to capture complex relationships) or overfitting (fitting noise rather than the true signal). Logistic Regression might have a lower AUC if the relationship between the features and the target is non-linear. Decision Tree could have a lowest AUC if it overfits or fails to generalize well to unseen data.

6 Conclusion

This chapter presents a comprehensive review of the application of machine learning techniques in predicting cardiovascular disease (CVD). Various methodologies, datasets, feature selection strategies, and performance evaluation metrics have been examined to understand the effectiveness of ML-based CVD prediction models. The findings emphasize the potential of machine learning, particularly ensemble methods, in enhancing accuracy and efficiency in CVD risk assessment. Despite the promising results, several challenges must be addressed to facilitate the widespread adoption of ML in clinical settings. These include data quality concerns, model interpretability, and ethical considerations. The chapter also analyzes prediction models to determine whether an individual is at risk of heart disease while providing insights for early diagnosis. By comparing performance metrics, we have observed that Gradient Boosting and Random Forest demonstrated the highest predictive accuracy compared to Decision Tree and Logistic Regression. Future research should focus on integrating multimodal data sources, developing interpretable machine learning models, improving model generalization, and conducting prospective validation studies across diverse patient populations to enhance the clinical applicability of ML-based CVD prediction.

Acknowledgements The work presented here is being conducted in the Department of Computer Science and Applications, MIT World Peace University, Pune, India. The authors would also like to thank Prof. Shubhalaxmi Joshi, Associate Dean, School of Computer Science and Technology for her kind support to carry out this work.

Authors Contribution Anu Singha- Conceptualization, Methodology, Data Curation, Formal Analysis, Visualization, Software, Validation, Writing—Original Draft, Proofreading, Supervision.

Mehul Bhatia- Conceptualization, Methodology, Visualization, Software, Writing—Original Draft.

Prajayshee Chauhan- Conceptualization, Methodology, Visualization, Software, Writing—Original Draft.

Sakshi Jaiswal- Conceptualization, Methodology, Visualization, Software, Writing—Original Draft.

Human Participants and/or Animals: N/A.

Funding There is no funding for this research.

Data Availability No/Not applicable (this manuscript does not report data generation or analysis).

Compliance with Ethical Standards

Conflict of Interest The author declares no potential conflict of interest with respect to the authorship and/or publication of this article.

Ethics Approval No, all of the material is owned by the authors and/or no permissions are required.

References

1. Weng SF, Reps J, Kai J, Garibaldi JM, Qureshi N (2017) Can machine-learning improve cardiovascular risk prediction using routine clinical data?" J PLoS One 12(4)
2. Saboor A, Muhammad U, Sikandar A, Samad A, Muhmmad FA, Najeeb U (2022) A method for improving prediction of human heart disease using machine learning algorithms. Mob Inf Syst 1410169:9 pp
3. Zaman MIU, Tabassum S, Ullah MS, Rahaman A, Nahar S, Islam AM (2019) Towards IoT and ML driven cardiac status prediction system. In: Proceedings in 1st international conference on advances in science, engineering and robotics technology (ICASERT), pp 1–6

4. Krittanawong C, Virk HUH et al (2020) Machine learning prediction in cardiovascular diseases: a meta-analysis. J Sci Rep 10
5. Bharti R, Khamparia A, Shabaz M, Dhiman G, Pande S, Singh P (2021) Prediction of heart disease using a combination of machine learning and deep learning. J Comput Intell Neurosci. Wiley
6. Ahmed H, Abd-El Ghany SF, Youn EMG, Omran NF, Ali AA (2019) Stroke prediction using distributed machine learning based on apache spark. Int J Adv Sci Technol 28(15):89–97
7. Biswas N, Uddin KMM, Rikta ST, Dey SK (2022) A comparative analysis of machine learning classifiers for stroke prediction: a predictive analytics approach. J Healthc Anal 2
8. Maini E, Venkateswarlu B, Gupta A (2018) Applying machine learning algorithms to develop a universal cardiovascular disease prediction system. In: International conference on intelligent data communication technologies and internet of things (ICICI). Springer
9. Enriko IKA, Suryanegara M, Gunawan DA (2016) Heart disease prediction system using K-nearest neighbor algorithm with simplified patient's health parameters. J Telecommun Electron Comput Eng
10. Anitha S, Sridevi N (2019) Heart disease prediction using data mining techniques. J Anal Comput 8(2):48–55
11. Hazra A, Mandal SK, Gupta A, Mukherjee A, Mukherjee A (2017) Heart disease diagnosis and prediction using machine learning and data mining techniques: a review", Adv Comput Sci Technol 10(7)
12. Hosmer D, Lemeshow S (1989) Applied logistic regression. Wiley, New York
13. Wu X et al (2008) Top 10 algorithms in data mining. Knowl Inf Syst 14:1–37
14. Breiman L (2001) Random forests. Mach Learn 45:5–32
15. Friedman JH (2001) Greedy function approximation: a gradient boosting machine. Ann Stat 1189–1232

Generative Models for Image Synthesis

Deepak Dhillon, Satya Prakash Yadav, and Aishwary Varshney

Abstract Generative models have become a cornerstone in the field of artificial intelligence, particularly for image synthesis, enabling the creation of high-quality, realistic images across various domains. This chapter provides a comprehensive overview of key generative models, including Generative Adversarial Networks (GANs), Variational Autoencoders (VAEs), Denoising Diffusion Probabilistic Models (DDPMs), and more. We delve into the architectures, sub-models, and unique capabilities of each, highlighting their applications in image generation, translation, and enhancement. Additionally, we discuss the challenges these models face, such as training instability and computational complexity, and explore future directions in generative image synthesis. This exploration provides a deep understanding of how these models are shaping the future of AI-driven creativity and design.

Keywords Generative models · Image synthesis · GANs · VAEs · Diffusion models

1 Introduction

1.1 Overview of Generative Models and Their Significance in Image Synthesis

Generative models are a class of machine learning models designed to generate new data instances that mimic the distribution of a given dataset. Unlike discriminative models, which focus on predicting labels or classes from data, generative models learn the underlying patterns, relationships, and structures within the data, allowing them to create entirely new instances. In the context of image synthesis,

D. Dhillon (✉)
School of Artificial Intelligence, Bennett University, Greater Noida, UP, India
e-mail: a24soaip0003@bennett.edu.in

S. P. Yadav · A. Varshney
School of Engineering and Information Technology, Sanskriti University, Mathura, UP, India

S. Chand Kumain et al. (eds.), *AI and ML Techniques in Image Processing and Object Detection*, https://doi.org/10.1007/978-981-96-7445-9_9

these models have revolutionized the way we generate, transform, and enhance images, providing new capabilities in fields ranging from art and entertainment to healthcare and defence.

The significance of generative models lies in their ability to perform complex image-related tasks with minimal human intervention. They can create photorealistic images, translate images between domains (such as turning sketches into fully coloured artworks), enhance low-resolution images, and even generate novel content based on textual descriptions. These capabilities have opened up new avenues for creativity and practical applications, making generative models a critical component of modern artificial intelligence.

1.2 Historical Background and Evolution of Generative Models

The journey of generative models began with basic probabilistic approaches like Gaussian Mixture Models (GMMs) and Hidden Markov Models (HMMs), which laid the groundwork for understanding data distribution. As computational power increased and neural networks evolved, the field experienced significant advancements, leading to the development of more sophisticated models.

Early neural network-based approaches, such as Restricted Boltzmann Machines (RBMs) and Deep Belief Networks (DBNs), provided the first glimpses of generative capabilities within neural architectures. However, the true breakthrough came with the introduction of Generative Adversarial Networks (GANs) by Ian Goodfellow and his colleagues in 2014. GANs introduced a novel adversarial training mechanism, where two neural networks—the generator and the discriminator—compete against each other, leading to highly realistic image generation.

Following GANs, Variational Autoencoders (VAEs) emerged as a powerful framework, utilizing probabilistic graphical models and deep learning to generate data with meaningful latent representations. VAEs offered more stable training compared to GANs and brought about new insights into how generative models could structure and interpret data.

More recently, models such as Denoising Diffusion Probabilistic Models (DDPMs), autoregressive models like PixelCNN and PixelRNN, and transformer-based architectures have pushed the boundaries of what is possible in image synthesis see Fig. 1. These models have achieved remarkable success in generating high-resolution images, integrating textual inputs, and synthesizing complex scenes with exceptional detail.

This chapter provides an in-depth exploration of key generative models used in image synthesis, such as GANs, VAEs, and DDPMs, along with their architectures, mechanisms, strengths, and limitations. It delves into various sub-models and their specific applications, from artistic image generation to practical uses in healthcare and industry. Additionally, the chapter addresses challenges like training instability, mode

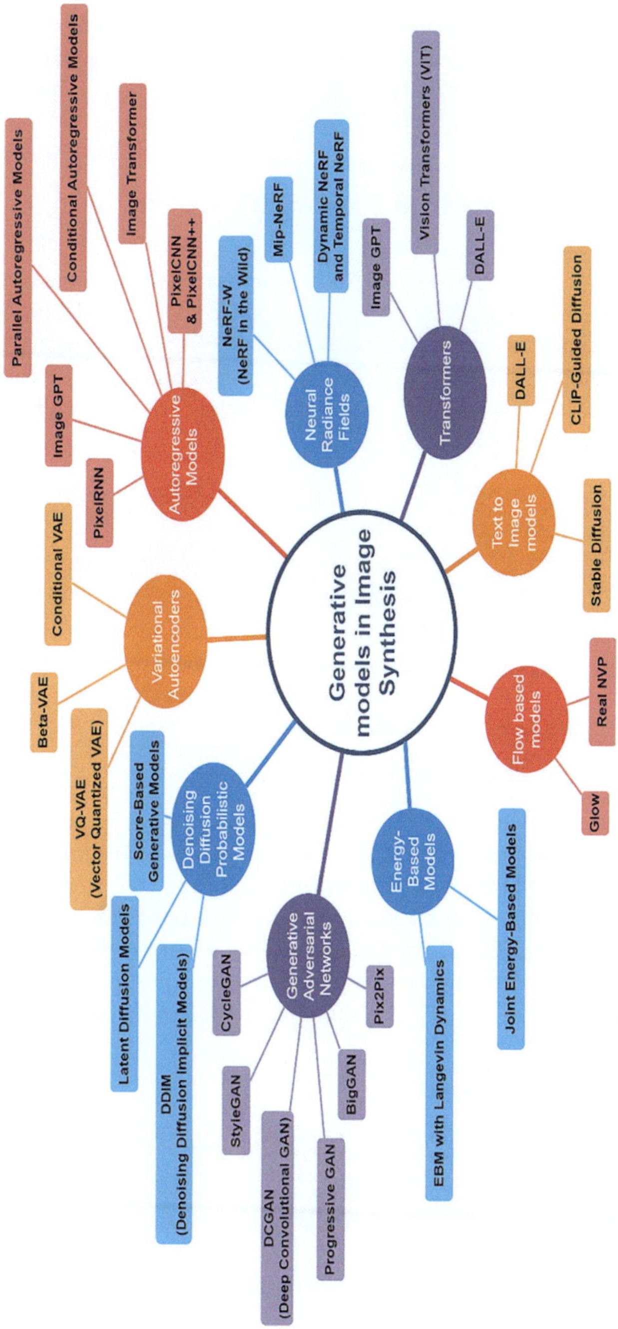

Fig. 1 Generative models in image synthesis

collapse, and ethical considerations, offering a balanced view of the current landscape and future potential of generative models in AI. Readers will gain a comprehensive understanding of these powerful tools and their broader implications.

2 Generative Adversarial Networks (GANs)

Generative Adversarial Networks (GANs) have become one of the most influential frameworks in the realm of image synthesis since their introduction in 2014. By leveraging a unique adversarial training process, GANs can generate images that are highly realistic, often indistinguishable from real images. This section explores the architecture, variants, applications, and challenges of GANs, highlighting their transformative impact on the field [1].

2.1 Architecture of GANs

At the core of GANs is the interplay between two neural networks: the Generator and the Discriminator. These networks are trained simultaneously in a zero-sum game where the Generator aims to produce realistic data, and the Discriminator strives to distinguish between real and generated data.

Generator and Discriminator Roles:

Generator (G): The Generator is responsible for creating synthetic data (images, in this case) from random noise. It takes a random vector (usually sampled from a Gaussian distribution) as input and transforms it through a series of neural network layers to produce a high-dimensional output that resembles the real data.

Discriminator (D): The Discriminator acts as a binary classifier that evaluates the authenticity of the images. It receives both real images from the training set and synthetic images generated by the Generator. Its goal is to correctly classify these images as either real or fake.

Adversarial Training Dynamics:

The adversarial process involves the Generator and Discriminator engaging in a continuous battle:

- The **Generator** attempts to fool the Discriminator by generating increasingly realistic images.
- The **Discriminator** tries to improve its classification abilities to detect synthetic images accurately.

This adversarial relationship is formalized through a minimax optimization problem, where the Generator aims to minimize the Discriminator's ability to distinguish real from fake, and the Discriminator tries to maximize its classification accuracy. Mathematically, the objective function is:

$$\min_{G}\max_{D} V(D, G) = \mathbb{E}_{z \sim P_{data}(x)}\left[logD(x)\right] + \mathbb{E}_{z \sim p_z(z)}\left[\log(1 - D(G(z)))\right]$$

where P_{data} (x) is the distribution of real data, and p_z(z) is the distribution of the noise vector z.

This adversarial training dynamic is the key innovation that drives GANs, enabling them to learn complex data distributions effectively.

2.2 *Sub-models and Variants*

Since their inception, GANs have evolved into numerous sub-models and variants, each designed to address specific challenges or expand the capabilities of the original framework.

DCGAN (Deep Convolutional GAN):

- Utilizes deep convolutional layers, making GANs more stable and suitable for image-related tasks [2].
- Key advancements include the removal of fully connected layers and the use of strided convolutions, resulting in higher quality images.

StyleGAN and StyleGAN2:

- Introduced a style-based architecture that disentangles high-level attributes (like pose) from fine details (like colour), allowing fine-grained control over the generated images [3].
- Widely used for generating highly realistic human faces and other complex textures.

CycleGAN:

- Designed for unpaired image-to-image translation tasks, such as converting photographs to paintings without paired training data.
- Uses cycle-consistency loss to ensure that image translations can be reversed accurately.

Pix2Pix:

- Focuses on paired image-to-image translation tasks, such as converting sketches to realistic images.
- Directly learns the mapping from input to output images with a conditional GAN approach.

BigGAN:

- Scales up GANs to handle high-resolution images by increasing model capacity and dataset size, producing images with unprecedented quality and diversity.

Progressive GAN:

- Introduces progressive training, starting with low-resolution images and gradually adding layers to increase resolution, improving training stability and output quality.

3 Variational Autoencoders (VAEs)

Variational Autoencoders (VAEs) represent a powerful and flexible framework for generative modelling, blending probabilistic graphical models with deep learning. Unlike traditional autoencoders that focus on dimensionality reduction and reconstruction, VAEs learn to encode data into a latent space that captures meaningful variations, enabling the generation of new, diverse data samples [4]. This section explores the architecture, sub-models, applications, and challenges of VAEs, highlighting their unique contributions to image synthesis.

3.1 VAE Architecture and Mechanisms

VAEs are a type of probabilistic generative model that combine deep learning with Bayesian inference. The architecture consists of two main components: the encoder, which maps input data to a latent space, and the decoder, which reconstructs the data from this latent representation.

Latent Space Representation and Probabilistic Approach:

Encoder: The encoder network maps the input image xxx to a distribution over the latent space, typically modelled as a multivariate Gaussian distribution. Instead of encoding the image into a single point, the encoder outputs the mean (μ) and standard deviation (σ) of the latent distribution, allowing for a probabilistic representation.

Latent Space: The latent space is a compressed, abstract representation of the data that captures underlying features and variations. By sampling from the latent distribution, the model can generate new, similar images, providing a versatile way to explore the data manifold.

Decoder: The decoder reconstructs the input by mapping the sampled latent vectors back to the image space. The process of decoding leverages the learned patterns in the latent space to produce high-quality reconstructions.

Probabilistic Approach: The key innovation of VAEs is their use of a probabilistic framework to learn the latent representation. The loss function consists of two parts: the reconstruction loss, which measures how well the decoder reconstructs the input, and the Kullback–Leibler (KL) divergence, which regularizes the latent space to follow a standard normal distribution. The objective is to minimize:

$$\mathfrak{L}_{VAE} = \mathbb{E}_{q(z|x)}[logp(x|z)] - D_{KL}(q(z|x)||p(z)||)$$

This combination of reconstruction and regularization allows the model to generate diverse and coherent samples from the latent space.

3.2 Sub-models and Variants

Since their introduction, VAEs have been adapted into various sub-models, each enhancing different aspects of the original architecture:

Beta-VAE:

- A modification of the standard VAE that introduces a scaling factor (β) to the KL divergence term. This adjustment allows for more explicit control over the trade-off between reconstruction quality and the disentanglement of latent factors, making it useful for learning interpretable representations.

Conditional VAE (CVAE):

- Extends the VAE by conditioning the generation process on additional information, such as labels or attributes. This makes CVAEs particularly suitable for tasks where image generation needs to be guided by specific characteristics, such as generating images of a certain class.

Vector Quantized VAE (VQ-VAE):

- Combines the strengths of VAEs and discrete latent variable models by incorporating vector quantization into the latent space. This approach discretizes the latent space, enhancing the model's ability to capture complex patterns and making it more suitable for tasks like image compression and high-fidelity generation.

4 Denoising Diffusion Probabilistic Models (DDPMs)

Denoising Diffusion Probabilistic Models (DDPMs) have emerged as a powerful class of generative models, offering an alternative approach to traditional frameworks like GANs and VAEs. DDPMs use a step-by-step denoising process to generate high-quality images, achieving impressive results in areas like inpainting and text-to-image synthesis [5]. This section delves into the mechanisms of diffusion models, their key variants, applications, and the challenges associated with their use.

4.1 Mechanism of Diffusion Models

Diffusion models are based on the idea of gradually transforming data into noise and then learning to reverse this process to reconstruct the data. This forward and reverse diffusion approach underpins their ability to generate images that closely resemble real data [6].

Forward and Reverse Diffusion Processes:

Forward Diffusion Process:

- In the forward diffusion process, the model gradually adds noise to an image over a sequence of time steps, resulting in a noisy, unstructured output. The objective is to corrupt the data distribution into a standard Gaussian distribution through a series of small perturbations.
- Mathematically, the forward process is defined as a Markov chain where each step adds a small amount of Gaussian noise, making the image progressively noisier until it resembles pure noise.

Reverse Diffusion Process:

- The reverse process aims to learn the step-by-step denoising of the noisy image back into its original form. This is achieved by training a neural network to predict the noise added at each step, effectively reversing the forward diffusion.
- The reverse process can be interpreted as sampling from a learned distribution, moving from random noise back to a high-quality, coherent image. The model is trained to minimize the difference between the predicted noise and the actual noise added during the forward process.

Probabilistic Modelling:

- The key innovation in diffusion models is the use of probability distributions to model each step of the forward and reverse processes. This probabilistic approach allows the model to generate diverse samples and effectively capture the underlying structure of the data.

4.2 Sub-models and Variants

Diffusion models have been adapted into various sub-models and variants, each introducing enhancements to improve efficiency, flexibility, or output quality.

Denoising Diffusion Implicit Models (DDIM):

- DDIM modifies the original DDPM framework by introducing a non-Markovian sampling process, allowing for faster and more efficient sampling. By adjusting the number of sampling steps, DDIM can balance between generation speed and quality, making it suitable for real-time applications.

Latent Diffusion Models (LDMs):

- LDMs incorporate latent representations, similar to VAEs, to perform diffusion in a compressed space rather than pixel space. This approach significantly reduces computational complexity and allows for handling higher resolutions and more complex data types, such as videos or 3D structures.

Score-Based Generative Models:

- These models, closely related to DDPMs, use a score-matching approach to learn the gradient of the data distribution. They can generate samples by following gradients of learned score functions, offering a continuous generalization of diffusion models.

5 Autoregressive Models

Autoregressive models are a class of generative models that generate images by sequentially predicting pixels or groups of pixels based on previous predictions. They are particularly known for their ability to capture intricate pixel-level details, making them effective for tasks requiring high-quality, coherent outputs. This section covers the architecture and mechanisms of autoregressive models, their applications, and the challenges involved in their use.

5.1 Architecture and Mechanisms

Autoregressive models generate images by modelling the probability distribution of pixel values in a sequential manner. Each pixel is conditioned on the previous pixels, creating a chain-like process where each step depends on the previous context.

PixelRNN:

- PixelRNN is one of the earliest autoregressive models designed for image generation. It uses recurrent neural networks (RNNs) to sequentially predict each pixel's value row-by-row or in a zigzag manner across the image. The model captures complex dependencies between pixels but is computationally expensive due to its sequential nature.

PixelCNN:

- PixelCNN builds on the idea of PixelRNN but replaces the RNN structure with convolutional layers. This modification allows the model to generate pixels in parallel within a single layer, significantly speeding up the generation process while maintaining the autoregressive property. Variants like Gated PixelCNN introduce gating mechanisms to enhance pixel dependencies [7].

Image GPT:

- Inspired by the success of GPT models in natural language processing, Image GPT applies a similar transformer-based autoregressive architecture to image generation. It treats images as sequences of pixels, using attention mechanisms to model the relationship between pixels over long distances. Image GPT is capable of generating coherent, high-quality images by learning complex dependencies across large datasets [8].

Mechanism of Autoregression:

- Autoregressive models predict each pixel's value conditioned on previously generated pixels, often using a likelihood-based approach. For an image xxx, the joint probability distribution is factorized as:

$$p(x) = p(x_1)p(x_2|x_1)p(x_3|x_1, x_2)\dots p(x_n|x_1, x_2, \dots, x_{n-1})$$

- This sequential prediction allows the model to generate detailed images that preserve local and global coherence.

5.2 *Sub-models and Variants*

PixelRNN (Recurrent Neural Networks)

- Uses recurrent neural networks to model the dependencies between pixels, generating images pixel by pixel.
- Captures long-range dependencies in images, useful for sequential generation.

PixelCNN and PixelCNN++

- A convolutional variant of PixelRNN, which uses convolutions to capture pixel dependencies instead of recurrent connections, making it more efficient.
- PixelCNN++ introduced improvements like down sampling layers and more sophisticated conditioning mechanisms to enhance image quality.

Image Transformer

- Adapts transformer architectures for autoregressive image generation, focusing on self-attention mechanisms instead of convolutions or recurrences.

- Scales well with large data, capturing long-range dependencies better than traditional CNNs.

Image GPT (Generative Pre-trained Transformer)

- A variant of GPT adapted for images, which treats image synthesis as a language modelling problem by predicting pixels or tokens sequentially.
- Leverages large-scale pre-training, self-attention, and autoregressive prediction to generate high-quality images.

Conditional Autoregressive Models

- Extends autoregressive models by conditioning the generation process on additional information, such as text descriptions or class labels.
- Enables targeted image generation based on specific conditions, enhancing control over the output.

Parallel Autoregressive Models

- Attempts to parallelize the generation process of traditional autoregressive models by predicting multiple pixels simultaneously while maintaining dependencies.
- Faster generation times compared to sequential models without sacrificing output quality.

6 Neural Radiance Fields (NeRF)

Neural Radiance Fields (NeRF) have revolutionized the field of 3D image synthesis by providing a novel approach to representing and rendering 3D scenes. By using neural networks to model scenes as continuous volumetric data, NeRF enables high-fidelity 3D reconstructions and realistic novel view synthesis from 2D images. This section explores the underlying mechanisms of NeRF, key variants, applications, and the challenges faced in optimizing these models [9].

6.1 NeRFs and Their Mechanisms

NeRF is a neural network-based approach that models 3D scenes by learning a continuous volumetric representation. Unlike traditional 3D modelling techniques that use explicit meshes or voxels, NeRF leverages neural networks to encode the colour and density of points in 3D space, allowing for high-quality rendering of complex scenes.

3D Scene Representation as Continuous Volumetric Data:

- NeRF models a 3D scene by mapping spatial coordinates (x, y, z) and viewing direction (θ, ϕ) to RGB color values and volumetric density. This mapping is achieved through a fully connected neural network that learns to encode the scene based on input images taken from different angles.

- During rendering, NeRF integrates the colour and density values along a ray cast through the scene to compute the final pixel value. This approach allows the model to synthesize novel views of the scene that were not part of the training data, creating highly realistic visual effects.
- The training process involves minimizing the difference between the rendered views and the actual input images, allowing the model to accurately capture the fine details and lighting conditions of the scene.

6.2 *Sub-models and Variants*

NeRF has inspired several variants that enhance its capabilities, address specific limitations, or optimize its performance for different applications.

Mip-NeRF:

- Mip-NeRF improves the efficiency and quality of NeRF by incorporating a multi-scale representation of the scene. By using mipmaps (precomputed, optimized sequences of images), Mip-NeRF effectively handles anti-aliasing and improves the rendering of fine details at varying distances.
- This variant allows for faster training and rendering, making it more suitable for real-time applications where performance is critical.

NeRF-W:

- NeRF-W extends NeRF to handle unstructured and variable lighting conditions, making it ideal for real-world scenarios where lighting can change dynamically. By introducing additional latent variables, NeRF-W learns to separate scene geometry from transient lighting effects, enabling robust rendering under diverse conditions.
- This model is particularly useful for environments where consistent lighting is not guaranteed, such as outdoor scenes or dynamic lighting in indoor spaces.

Dynamic NeRF and Temporal NeRF:

- These models extend the original NeRF framework to capture dynamic scenes, including moving objects and temporal changes. By incorporating time as an additional input, these variants can generate animations and realistic transitions, broadening the scope of NeRF's applications.

7 Transformers in Image Synthesis

Transformers have revolutionized various domains of machine learning, including image synthesis. Their ability to handle long-range dependencies and process sequences has extended to the generation and transformation of images. This section explores transformer architectures used in image synthesis, their applications, and the challenges they face [10].

7.1 Transformer Architectures

Transformers, initially designed for natural language processing, have been adapted for image synthesis with remarkable success. Key transformer-based models in this field include Vision Transformers, Image GPT, and DALL-E.

Vision Transformers (ViTs):

- Vision Transformers adapt the transformer architecture, which was originally designed for sequences of words, to handle images by treating them as sequences of patches. Instead of processing individual pixels, ViTs divide an image into fixed-size patches, which are then linearly embedded into a sequence.
  - The image patches are flattened and linearly projected into embeddings. These embeddings are then fed into a transformer encoder, which processes the sequence using self-attention mechanisms.
  - The transformer architecture learns global relationships and contextual information across the entire image, allowing it to capture complex patterns and details.

Image GPT:

- Image GPT is an adaptation of the GPT architecture for image generation. It treats images as sequences of pixels or patches and uses a transformer decoder to generate images autoregressively [8].
  - Image GPT uses a masked language modelling approach, where parts of the image are masked during training, and the model learns to predict the missing pixels based on the context provided by the visible parts.
  - The model generates images pixel by pixel or patch by patch, capturing high-level patterns and details through the transformer's attention mechanisms.

DALL-E:

- DALL-E extends the transformer approach to generate images from textual descriptions. It combines a transformer-based text encoder with an image decoder to create visual content from textual input [10].

- DALL-E uses a VQ-VAE-2 (Vector Quantized Variational Autoencoder) to represent images in a discrete latent space, where a transformer model learns to map textual descriptions to image tokens.
- The model generates images by autoregressively predicting sequences of image tokens based on the input text.

8 Text-To-Image Models

Text-to-image models have transformed the way we generate visual content from textual descriptions, enabling creative and practical applications across various fields. These models leverage advances in machine learning to create detailed images based on textual input, bridging the gap between language and vision [11]. This section explores the key models in text-to-image generation, their applications, and the challenges they face.

8.1 Models and Mechanisms

Text-to-image models use advanced neural networks to generate images from textual descriptions. Several notable models in this area include DALL-E, CLIP-Guided Diffusion, and Stable Diffusion.

DALL-E:

- Developed by OpenAI, DALL-E is a groundbreaking model that generates images from textual prompts using a combination of a transformer-based text encoder and a VQ-VAE-2 image decoder.
  - DALL-E uses a discrete latent space to represent images and learns to map textual descriptions to sequences of image tokens. The model generates images by autoregressively predicting these tokens based on the input text.
  - It excels at creating novel and imaginative visual content, combining elements in ways that may not exist in the real world.

Capabilities:

- DALL-E can generate diverse and complex images, including abstract concepts and creative visual combinations, making it suitable for art, design, and entertainment.

CLIP-Guided Diffusion:

- CLIP-Guided Diffusion combines the capabilities of OpenAI's CLIP (Contrastive Language–Image Pre-training) model with diffusion-based image generation techniques.

- CLIP is used to evaluate and guide the diffusion process, which iteratively refines a noisy image into a coherent one based on the textual prompt. CLIP's role is to ensure that the generated image aligns with the text description.
- This approach leverages CLIP's ability to understand and match text-image relationships to produce high-quality, contextually relevant images.

Capabilities:

- CLIP-Guided Diffusion is particularly effective at generating images that closely match the textual description, even in complex or nuanced scenarios.

Stable Diffusion:

- Stable Diffusion is a model that generates high-quality images by learning a stable latent space for image synthesis. It integrates diffusion techniques with advanced generative models.
  - The model uses a diffusion process to iteratively refine an image from noise based on textual input, guided by a latent space that captures diverse image features and contexts.
  - This approach balances image quality and stability, resulting in high-resolution images with consistent detail and coherence.

Capabilities:

- Stable Diffusion is known for producing stable, high-quality images while maintaining interpretability and control over the generation process.

9 Flow-Based and Energy-Based Models

Flow-based and energy-based models are powerful approaches in generative modelling, each offering unique advantages for image synthesis and complex distribution modelling. This section delves into these models, their mechanisms, and their applications.

9.1 Flow-Based Models

Flow-based models learn to transform a simple distribution into a complex one using a series of invertible transformations. They are characterized by their ability to provide exact likelihood estimates and perform exact sampling.

Real NVP (Real-Valued Non-Volume Preserving):

- Real NVP is a flow-based model that uses a series of coupling layers to transform a simple distribution into a complex one. The model applies a series of bijective (invertible) transformations, allowing for exact likelihood computation and sampling.
- The transformation is split into two parts: a coupling layer that updates a subset of the variables while keeping the rest fixed, and an affine transformation that ensures invertibility.

Glow:

- Glow extends Real NVP by incorporating additional features, such as 1×1 convolutions and invertible residual networks, to improve modelling capacity and flexibility.
- Glow uses a series of invertible layers to transform data, with the key feature being the use of reversible 1×1 convolutions to model the dependencies between variables more effectively.

9.2 Energy-Based Models

Energy-based models (EBMs) are characterized by their use of an energy function to model data distributions. These models aim to learn an energy function that assigns low energy to data samples and high energy to non-samples.

EBM with Langevin Dynamics:

- EBMs use Langevin Dynamics to sample from the learned distribution by iteratively updating samples based on the gradient of the energy function. This process helps in generating samples that approximate the target distribution.
- Langevin Dynamics involves adding noise to the samples and moving them in the direction of decreasing energy, thereby converging to regions of low energy.

Joint Energy-Based Models:

- Joint EBMs model the joint distribution of multiple variables by learning an energy function over the combined space. This approach allows for modelling complex dependencies between variables.
- These models use energy functions to capture relationships between variables, providing a unified framework for understanding and generating multi-dimensional data.

10 Applications and Use Cases

Table 1 outlines the primary applications and use cases for each generative model and its variants.

Table 1 Applications and use cases of generative models of image synthesis

Model type	Sub-models	Applications and use cases
Generative Adversarial Networks (GANs)	– DCGAN – StyleGAN – CycleGAN – Pix2Pix – BigGAN – Progressive GAN	– DCGAN: realistic image generation, unsupervised feature learning – StyleGAN: high-resolution image synthesis, portrait generation – CycleGAN: image-to-image translation, domain adaptation – Pix2Pix: image-to-image translation, data augmentation – BigGAN: large-scale image synthesis, generating high-quality images – Progressive GAN: high-resolution image generation, detailed texture synthesis
Variational Autoencoders (VAEs)	– Beta-VAE – Conditional VAE – VQ-VAE	– Beta-VAE: image reconstruction, disentangled representation learning – Conditional VAE: semi-supervised learning, conditional image generation – VQ-VAE: high-quality image reconstruction, speech synthesis
Denoising Diffusion Probabilistic Models (DDPMs)	– DDIM – Latent diffusion models	– DDIM: high-quality image synthesis, denoising tasks – Latent diffusion models: text-to-image generation, image inpainting
Autoregressive models	– PixelRNN – PixelCNN – Image GPT	– PixelRNN: sequential image generation, high-resolution image synthesis – PixelCNN: image generation, inpainting, and super-resolution – Image GPT: high-resolution image generation, text-to-image synthesis

(continued)

Table 1 (continued)

Model type	Sub-models	Applications and use cases
Neural Radiance Fields (NeRF)	– Mip-NeRF – NeRF-W	– Mip-NeRF: 3D scene reconstruction, detailed scene rendering – NeRF-W: view synthesis for complex scenes, virtual reality (VR) content creation
Transformers in image synthesis	– Vision Transformers – Image GPT – DALL-E	– Vision transformers: high-resolution image generation, image classification – Image GPT: high-resolution image generation, text-to-image synthesis – DALL-E: creative content generation, text-to-image synthesis
Text-to-image models	– DALL-E – CLIP-guided diffusion – Stable diffusion	– DALL-E: creative content generation, concept visualization – CLIP-guided diffusion: text-to-image synthesis, creative content generation – Stable diffusion: high-quality text-to-image generation, artistic image synthesis
Flow-based models	– Real NVP – Glow	– Real NVP: exact likelihood estimation, image generation, density estimation – Glow: high-quality image generation, complex distribution modelling
Energy-based models	– EBM with Langevin dynamics – Joint energy-based models	– EBM with Langevin dynamics: sampling from complex distributions, image synthesis – Joint energy-based models: multi-modal data generation, complex distribution modelling

11 Challenges and Future Directions

Generative models for image synthesis have made remarkable advancements, yet several challenges persist that need addressing to unlock their full potential. Training stability remains a significant issue across various model types, with instability and mode collapse affecting the quality of generated outputs. Additionally, balancing quality and diversity in generated images presents a trade-off, as models often struggle to maintain high fidelity while producing a wide range of outputs. Ethical considerations are also critical, as the potential for misuse—such as generating misleading or harmful content—necessitates careful management and regulation. Furthermore, the computational demands of advanced models can be prohibitive, limiting accessibility and practical use. Addressing these challenges involves developing more robust training methods, implementing ethical guidelines, and optimizing models to reduce computational requirements, ultimately making these technologies more accessible and effective. Table 2 provides a consolidated view of the challenges and solutions for each type of generative model discussed in the chapter.

Table 2 Challenges and solutions of generative models of image synthesis

Model type	Challenges	Solutions
Generative Adversarial Networks (GANs)	– Training instability and mode collapse – High computational cost – Difficulty in evaluating model performance	– Use of Wasserstein GANs (WGAN) for stability – Spectral normalization to stabilize training – Advanced architectures like Progressive GANs to improve training – Metrics like Inception Score (IS) and Fréchet Inception Distance (FID) for evaluation
Variational Autoencoders (VAEs)	– Blurrier outputs compared to GANs – Balancing reconstruction and regularization – Difficulty in capturing complex data distributions	– Use of improved VAE variants (e.g., Beta-VAE) to enhance image quality – Incorporate better regularization techniques to balance reconstruction and latent space – Use of normalizing flows or other enhancements to improve distribution modelling
Denoising Diffusion Probabilistic Models (DDPMs)	– High computational cost for diffusion steps – Long generation times – Challenges with high-dimensional data	– Use of more efficient diffusion processes (e.g., DDIM) to speed up sampling – Optimizations in diffusion algorithms to reduce computational overhead – Implementation of multi-scale or hierarchical approaches to handle high-dimensional data
Autoregressive models	– High computational intensity for pixel-level generation – Slow inference times – Difficulty in modelling long-range dependencies	– Use of efficient architectures (e.g., PixelSNAIL) to improve speed – Techniques like distillation to accelerate inference – Implement sparse attention mechanisms to better handle long-range dependencies

(continued)

Table 2 (continued)

Model type	Challenges	Solutions
Neural Radiance Fields (NeRF)	– High computational cost for training and rendering – Handling complex scenes and long rendering times – Difficulty in rendering fine details in large scenes	– Use of optimized NeRF variants (e.g., Mip-NeRF) to improve efficiency – Hierarchical approaches to manage scene complexity – Incorporate techniques for detail enhancement and faster rendering
Transformers in image synthesis	– Scalability issues with large images – Handling long-range dependencies – High computational and memory requirements	– Implement efficient transformer architectures (e.g., Swin Transformer) – Use of sparse attention mechanisms to manage large images – Optimize training and inference processes to reduce resource usage
Text-to-image models	– Biases in generated content – Ensuring alignment with complex textual descriptions – Handling diverse and abstract text prompts	– Implement bias mitigation techniques and regular audits – Use iterative refinement and enhanced training data for better alignment with text – Incorporate multi-modal learning to better handle diverse text prompts
Flow-based models	– Limited scalability to very high-dimensional data – High computational cost for training and inference – Complexity in managing the invertibility of transformations	– Utilize more scalable flow architectures (e.g., Glow with 1 × 1 convolutions) – Optimize training processes and hardware utilization – Develop advanced techniques to manage and verify the invertibility of transformations
Energy-based models	– Slow sampling with Langevin dynamics – Complexity in modelling high-dimensional joint distributions – Difficulty in training the energy function effectively	– Use faster sampling techniques or approximate methods – Leverage advanced training and optimization strategies for high-dimensional models – Implement regularization and training enhancements to improve the energy function

12 Conclusion

In this chapter, we explored the diverse landscape of generative models used in image synthesis, delving into key architectures such as GANs, VAEs, DDPMs, autoregressive models, and others. Each model offers unique mechanisms for generating high-quality, realistic images, with applications ranging from creative content generation to scientific simulations. We also highlighted their sub-models, variants, and real-world use cases across various industries.

Generative models have revolutionized image synthesis, pushing the boundaries of AI in creating lifelike images, improving design processes, and enhancing creative tasks. As these models continue to evolve, they hold transformative potential in areas such as virtual reality, healthcare, and even personalized digital experiences.

Looking ahead, ongoing advancements in model optimization, scalability, and accessibility will play a critical role in shaping the future of image synthesis. As we address the current challenges, including computational efficiency and ethical considerations, the potential of generative models in defining future technological landscapes remains vast and promising.

Acknowledgements The authors would like to express their gratitude to the researchers in the field of generative image processing, whose pioneering work has provided a strong foundation and inspiration for this study.

Author Contributions

- **Author 1**: Conducted the initial research and conceptualized the study.
- **Author 2**: Filtered the research findings and contributed to the writing process.
- **Author 3**: Generated images and performed final editing of the manuscript.

Funding This research did not receive any specific grant from funding agencies in the public, commercial, or not-for-profit sectors.

Data Availability All data and images used in this chapter have been generated by the authors and taken data is cited accordingly.

Conflict of Interest Statement The authors declare no conflict of interest.

References

1. Goodfellow IJ, Pouget-Abadie J, Mirza M, Xu B, Warde-Farley D, Ozair S, Courville A, Bengio Y (2014) Generative adversarial networks, 10 June 2014. https://arxiv.org/abs/1406.2661
2. Radford A, Metz L, Chintala S (2015) Unsupervised representation learning with deep convolutional generative adversarial networks, 19 November 2015. https://arxiv.org/abs/1511.06434
3. Karras T, Laine S, Aila T (2018) A style-based generator architecture for generative adversarial networks, 12 December 2018. https://arxiv.org/abs/1812.04948
4. Kingma DP, Welling M (2013) Auto-encoding variational Bayes, 20 December 2013. https://arxiv.org/abs/1312.6114
5. Ho J, Jain A, Abbeel P (2020) Denoising diffusion probabilistic models, 19 June 2020. https://arxiv.org/abs/2006.11239
6. Sohl-Dickstein J, Weiss EA, Maheswaranathan N, Ganguli S (2015) Deep unsupervised learning using nonequilibrium thermodynamics, 12 March 2015. https://arxiv.org/abs/1503.03585

7. Van Den Oord A, Kalchbrenner N, Vinyals O, Espeholt L, Graves A, Kavukcuoglu K (2016) Conditional image generation with PixelCNN decoders, 16 June 2016. https://arxiv.org/abs/1606.05328
8. Chen M, Radford A, Child R, Wu J, Jun H, Luan D, Sutskever I (2020) Generative pretraining from pixels. In: Proceedings of the 37th international conference on machine learning in proceedings of machine learning research, vol 119, pp 1691–1703. https://proceedings.mlr.press/v119/chen20s.html
9. Verbin D, Hedman P, Mildenhall B, Zickler TE, Barron JT, Srinivasan PP (2021) Ref-NeRF: structured view-dependent appearance for neural radiance fields. In: 2022 IEEE/CVF conference on computer vision and pattern recognition (CVPR), pp 5481–5490
10. Esser P, Rombach R, Ommer B (2020) Taming transformers for high-resolution image synthesis. In: 2021 IEEE/CVF conference on computer vision and pattern recognition (CVPR), pp 12868–12878
11. Gal R, Alaluf Y, Atzmon Y, Patashnik O, Bermano AH, Chechik G, Cohen-Or D (2022) An image is worth one word: personalizing text-to-image generation using textual inversion.arXiv: 2208.01618

Multi-class Classification of the SFM-Mass Images Using DL-Models with Machine Learning Classifiers

Jyoti Rani, Jaswinder Singh, and Jitendra Virmani

Abstract Breast cancer has emerged as a big reason of deaths in-between women. The multi-class classification of breast-masses using the SFM mass images is conducted by various computerized methods since last many years. Present work proposes a CAD design for multi-class classification of SFM mass images using deep feature-set and machine learning classifiers. The exhaustive experimentation is conducted by employing nine DL-based models and three ML based classifiers. These DL-based models used for extraction of deep feature-set are simple convolution series models / simple convolution DAG model/dilated convolution DAG models. The three ML-based classifiers i.e. ANFC-LH/ PCA-SVM/GA-SVM have been used extensively for classification task. Experimental work is carried on 518 SFM mass images chosen from DDSM dataset with 208 ϵ BIRAD-3, 150 ϵ BIRAD-4 and 160 ϵ BIRAD-5 classes, respectively. For segmenting masses from SFM mass images, ResNet50 semantic segmentation-model has been used. Segmented mass images are then used for extraction of deep feature-sets. The performance comparison of these DL-models, reports VGG19 model as the optimal model for deep feature-extractor. Deep feature-set is obtained using optimal feature extractor VGG19 model which may contain redundant values; therefore correlation based feature selection is employed to extract reduced deep feature-set. The performance of reduced deep feature-set is analyzed for multi-class classification using ML-based classifiers ANFC-LH, PCA-SVM and GA-SVM. The objective analysis of these CADs yields, VGG19 with ANFC-LH having highest estimated classification accuracy of 86% with individual class accuracy of 98, 80, 76% for BIRAD-3, BIRAD-4 and BIRAD-5 classes, respectively.

Keywords Mammograms · BIRAD classification · SFM mass images · MobileNet-V2 · VGG19 · ShuffleNet

J. Rani (✉)
GZSCCET, MRSPTU, Bathinda, India
e-mail: csejyotigill@gmail.com

J. Singh
Punjabi University, Patiala, India

J. Virmani
CSIR—CSIO, Chandigarh, India

S. Chand Kumain et al. (eds.), *AI and ML Techniques in Image Processing and Object Detection*, https://doi.org/10.1007/978-981-96-7445-9_10

1 Introduction

Mammography is the best technique often used for the breast cancer identification in the women with age having 38 years or may be above [1, 28, 46, 47, 56]. This imaging tool is capable of detecting breast cancer at the very early stage [8, 44, 54]. For the multi-class classification of cancerous or non cancerous masses using SFM mass images, a large number of computerized methods are being developed since last many years [4, 18, 20, 36, 39]. The mammographic mass classification methods have used traditional machine learning (ML) approaches by employing feature extraction [14], feature selection [43, 78] and classification [63–65, 67, 69, 76–78]. Later on with increasing performance as well as huge data processing capabilities, deep learning based CADs have been experimented on [25, 26, 30, 32, 37, 42, 48, 66]. Due to significant performance and improved accuracy, the DL-based CADs designs play an important role for classification of mammograms [2, 9, 15, 16, 20, 23, 24, 27].

Inspired from the deep network architectures, enhanced efficiency with fewer computations [37, 38, 45, 70, 74, 75], the present research work has extensively used nine deep learning based models belonging to different categories for multi-class classification of mammographic mass images into BIRAD-3, BIRAD-4 and BIRAD-5 classes, respectively [48]. The characterization performance of these nine DL based models for binary classification i.e. B3 (probably benign) and suspicious abnormality (B4 and B5 together considered as a single class) has been thoroughly investigated in the study [48]. However it is worth mentioning that differential diagnosis between suspicious abnormality as suspicious malignant (B4) and highly malignant (B5) is significantly important for prognosis. The present research work incorporates exhaustive experimentation for multi-class classification of the SFM mass images into B3, B4 and B5 (Probably benign, Suspicious malignancy and Highly Malignant classes). Characterization of the CADs for multi-class classification with SFM images is depicted in the Fig. 1.

The sample SFM mass images having BIRAD-3 (benign), BIRAD-4 (suspicious-malignant) and BIRAD-5 (highly-malignant) classes taken from DDSM data are presented in the Fig. 2.

2 Related Literature Work

The literature review reflects that large no. of work was conducted using DL models for the multi-class classification of the SFM mass images having both the original-images and the preprocessed-images [39, 52, 53, 57, 59–61, 72, 76]. The review of different work conducted for multi-class classification of original and the preprocessed SFM mass images with DDSM dataset has been presented here.

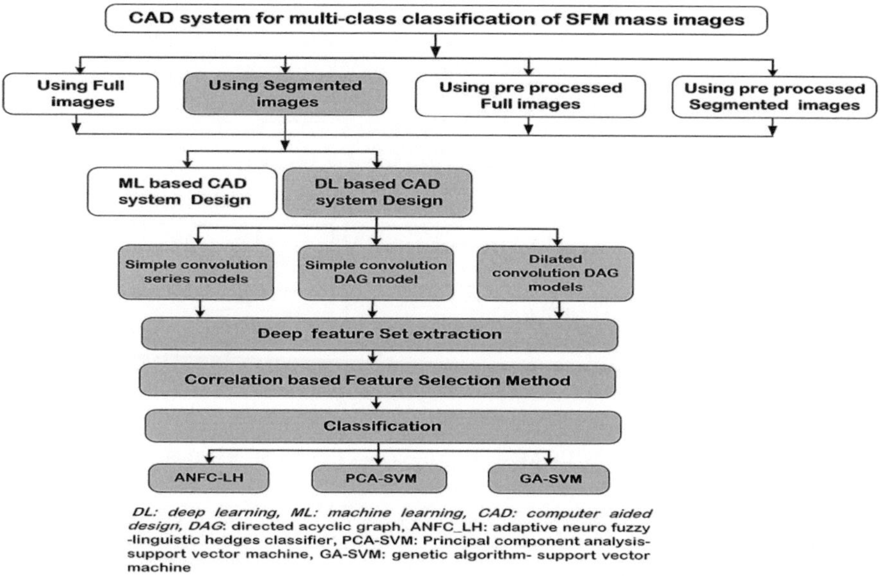

Fig. 1 Characterization of CADs for multi-class classification. ▇CAD systems used in the present work

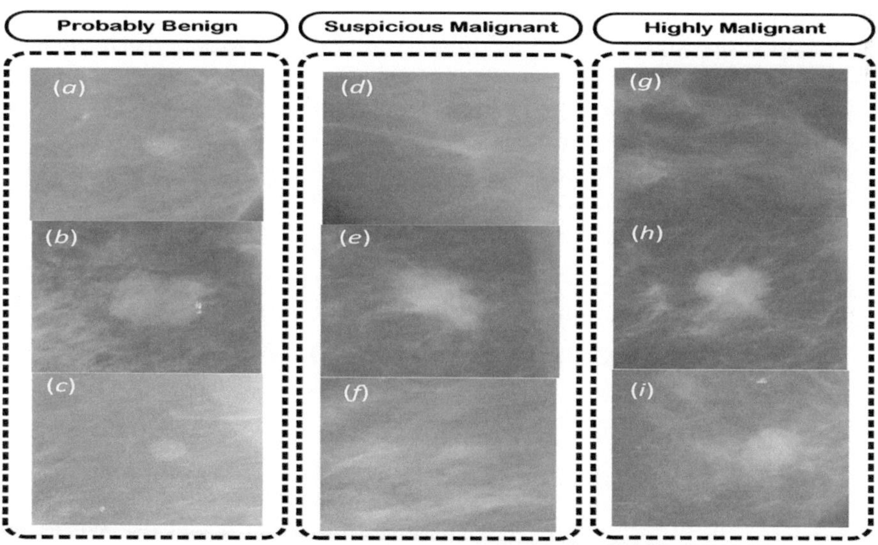

Fig. 2 Sample SFM mass images **a–c** BIRAD-3 *(probably-benign)* **d–f** BIRAD-4 *(suspicious-malignant)* **g–h** BIRAD-5 *(highly-malignant)*

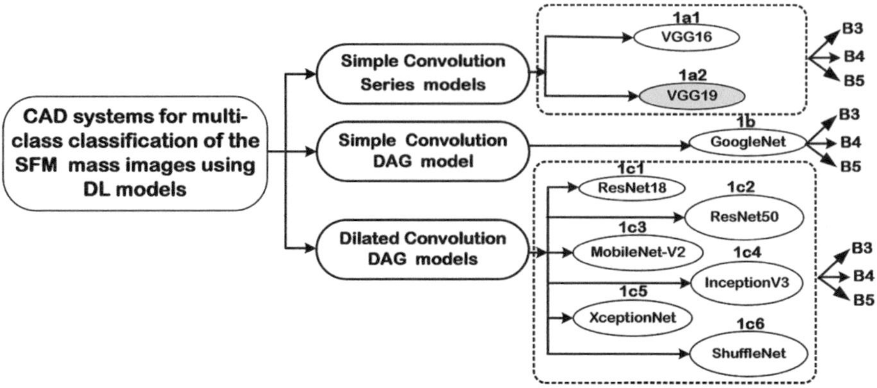

Fig. 3 CAD systems for multi-class classification of the SFM mass images using DL models. *Note DAG*: directed acyclic graph, *B3*: BIRAD-3, *B4*: suspicious malignant, *B5*: highly malignant

2.1 CAD Systems for Multi-Class Classification Using DL Models with SFM Mass Images

In Fig. 3, the characterization of CAD system using DL-based models used for multi-class classification of SFM mass images i.e. BIRAD-3 (benign), BIRAD-4 (suspicious-malignant) and BIRAD-5 (highly-malignant) classes, respectively have been presented.

Table 1 presents the review of work undertaken for multi-class classification of original SFM mass images on DDSM dataset.

Table 2 presents the work undertaken for multi-class classification using preprocessed SFM mass images on DDSM dataset.

2.2 CADs for Multi-class Classification of SFM Mass Images with Optimal Feature-Extractor and Machine Learning Based Classifiers

In Fig. 4, CAD for multi-class classification of the SFM mass images i.e. BIRAD-3 (benign), BIRAD-4 (suspicious-malignant) and BIRAD-5 (highly-malignant) classes respectively using optimal feature-extractor VGG19 and the ML-based classifiers are presented.

Table 3 presents studies employed with DL-models as feature-set-extractor/ML-classifier for classification using multi-class with original/preprocessed SFM mass images with DDSM data.

Table 1 The review of the work undertaken for the multi-class classification of the original SFM mass images on DDSM dataset

Author [Year]	No. of images	CNN model	Image classes	Image type	Evaluation parameters
Ballin et al. [8]	850	Fast RCNN	B1, B2, B3, B4, B5	Segmented	Accuracy-0.78, 0.77%
Samala et al. [55]	322	DCNN	B, M	Segmented	AUC-0.82 ± 0.02
Guan et al. [18]	2620	VGG16	N, Ab, B, M	Segmented	AUC-0.971
Yang et al. [73]	10,480	CNN	B, C	Segmented	Accuracy-92.31%
Ribli et al. [47]	2620	VGG 16	N, B, M	Full	AUC-0.85
Tariq et al. [64]	1586	VGG16, GoogleNet, InceptionV3	N, B, M	Segmented	Accuracy-81%
Sun et al. [58]	1445	MVDCNN	B, M	Segmented	Accuracy-81%
Tang et al. [66]	10,498	FCN	N, B, M	Segmented	AUC-81.37
Ubeyli [68]	2620	ResNet-150	N, B, M	Full	AUC-0.86
Li et al. [35]	2620	ResNet	B, M	Segmented	Accuracy-94.7%
Rani et al. [48]	518	VGG16/19, ResNet18/50, ShuffleNet, XceptionNet, MobileNetV2, GoogleNet	B, M	Segmented	Accuracy-96%

Note AUC: Area Under curve, *RCNN*: Region based convolutional Neural Network, *B1*: BIRAD-1, *B2*: BIRAD-2, *B3*: BIRAD-3, *B4*: BIRAD-4, *B5*: BIRAD-5, *N*: Normal, *B*: Benign, *Ab*: Abnormal, *M*: Malignant, D*CNN*: Deep Convolutional Neural Network

From the Tables 1, 2 to 3, it is evident that CAD systems for differential diagnosis between B3, B4 and B5 have not been experimented yet, however it is worth mentioning that with the progression of the malignant changes in the breast masses, the accurate differential diagnosis between B3, B4 and B5 classes by visual inspection becomes a daunting task for radiologists, motivated from this fact the present study has been carried out for optimal CAD system design for SFM mass classification into B3, B4 and B5 classes.

Table 2 The review of the work undertaken for the multi-class classification of the preprocessed SFM mass images on DDSM dataset

Author [Year]	Images	Preprocessed method	CNN	Image classes	Image type	Evaluation parameters
Jiao et al. [29]	800	Whitened	CNN	B, M	Segmented	Accuracy-96.8%
Carneiro et al. [11]	680	Gaussian filter	AlexNet	B, M	Segmented	VUS-0.9, AUC-0.9
Jadoon et al. [28]	2576	CLAHE	CNN	N, B, M	Full	Accuracy-79.92%
Debelee et al. [14]	2620	Histogram eqn	CNN	N, Ab	Segmented	Accuracy-98.9%
Aboutalib et al. [5]	9648	Histogram eqn	CNN	M, N, RB	Full	AUC-0.77–0.96
Abdelhafiz et al. [6]	2734	CLAHE	RU-Net, vanilla U-Net	B, M	Segmented	Accuracy-0.94%, Accuracy-0.93%
Nagaraj et al. [43]	2620	Histogram eqn	VGG16	B1, B2, B3, B4, B5	Full	Accuracy-83%
Gananasekaran et al. [21]	1416	CLAHE	VGG16	N, B, M	Segmented	Accuracy-96.47%, AUC-0.96
Yang et al. [74]	5706	Gaussian	DenseNet	B, M	Full	AUC-95.03, Accuracy-85%

Note RU-Net: Residual attention U-Net, *CLAHE*: Contrast Limited Adaptive Histogram Equalization, *Eqn*-Equalization, *Ng*: Negative, *RB*: Recalled benign, *B1*: BIRAD-1, *B2*: BIRAD-2, *B3*: BIRAD-3, *B4*: BIRAD-4, *B5*: BIRAD-5, *CNN*: Convolutional Neural Network

3 Experimental Methodology

The experimental work flow methodology adopted for present study is mentioned in the Fig. 5.

3.1 Dataset Description

The present experimental work is carried out on the public benchmark dataset known as Digital Database for Screening Mammography [25]. In present work, 518 mammographic mass images have been used. The dataset consists of 208 mammograms ϵ BIRAD-3, 150 mammograms ϵ BIRAD-4 and 160 mammograms ϵ BIRAD-5 classes respectively The dataset selection protocols have been formulated in consultation with participating radiologist, and accordingly SFM mass images with variable background density such as mild, moderate and severe densities and masses with round/

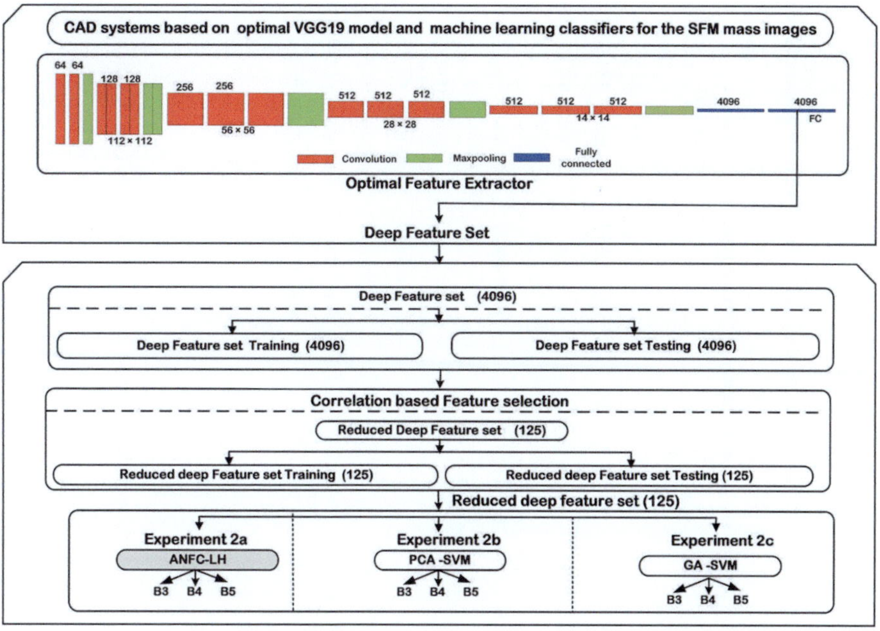

Fig. 4 CAD system based on optimal VGG19 model and machine learning based classifiers for SFM mass images. *Note B3*: BIRAD-3, *B4*: Suspicious malignant, *B5*: Highly malignant

oval and lobulated regular margins for BIRAD-3 class, with spiculated and irregular margins for BIRAD-4 class, with highly ill-defined and spiculated margins for BIRAD-5 class, with presence of micro/macro calcifications were selected for the present work.

The dataset preparation steps include (*a*) ROI cropping and (*b*) image resizing while preserving the aspect ratio [18, 25, 41, 48]. After the image resizing, the binary mask images are generated [48]. The binary mask images have been generated from cropped and resized images. The 518 SFM mass images are divided into training/testing dataset.

The SFM mass image dataset of 518 images belonging to BIRAD-3 class, BIRAD-4 class and BIRAD-5 class is not balanced as shown in Fig. 5. To make the balanced dataset for DL-based model training, the data augmentation has been incorporated using translations, rotations, horizontal/vertical flip operations [19, 48, 71]. For the details of these geometric transformations used for data augmentation the readers are directed to [48]. After data augmentation 10,864 segmented images and corresponding 10,864 mask images are fed to DL models. Dataset description is shown in Fig. 6.

Table 3 The review of studies employed with DL-model as feature-set-extractor/ML-classifier for classification using multi-class with original/ preprocessed SFM mass images with DDSM data

Author [Year]	No. of images	Original/Preprocessed images	DL-model	ML-classifiers	Image classes	Image type	Evaluation parameters
Jadoon et al. [28]	150	Pre-processed	CNN	SVM	N, B, M	Full	Accuracy 81.83–83.74%
Debelee et al. [14]	2620	Pre-processed	CNN	KNN	N, Ab	Segmented	Accuracy 98.9%
Al-antari et al. [3]	125	Pre-processed	DBN	LDA, NN	N, B, M		Accuracy 82.14, 82.57%
Song et al. [49]	11,562	Original	DLCNN	SVM XGBoost	N, B, C	Segmented	Accuracy 84, 92.8%
Malebary et al. [41]	1445	Original	CNN	RF	N, B, M	Full	Accuracy 0.96%
Dabbas et al. [16]	895	Original	AlexNet	HT	N, B, M	Segmented	Accuracy 92.10%
Sandler et al. [62]	1000	Original	AlexRes-Net	SVM-RBF	N, C	Segmented	Accuracy 95.87%
Kavitha et al. [31]	2500	Pre-processed	DLCN	BPNN	B, M	Segmented	Accuracy 97.55%
Rani et al. [48]	518	Original	VGG16/19, ResNet18/50, ShuffleNet, XceptionNet, MobileNetV2, GoogleNet	ANFC-LH, PCA-SVM, GA-SVM	B, M	Segmented	Accuracy 96%

Note XGBoost: Extreme gradient boosting, HT: Hanman Transform, SVM-RBF: Support Vector Machine-Radial Basis Function, *DLCN*: Deep Learning capsule networks, *BPNN*: Back propagation neural network, *N*: Normal, *C*: Cancer, RF: Random Forest

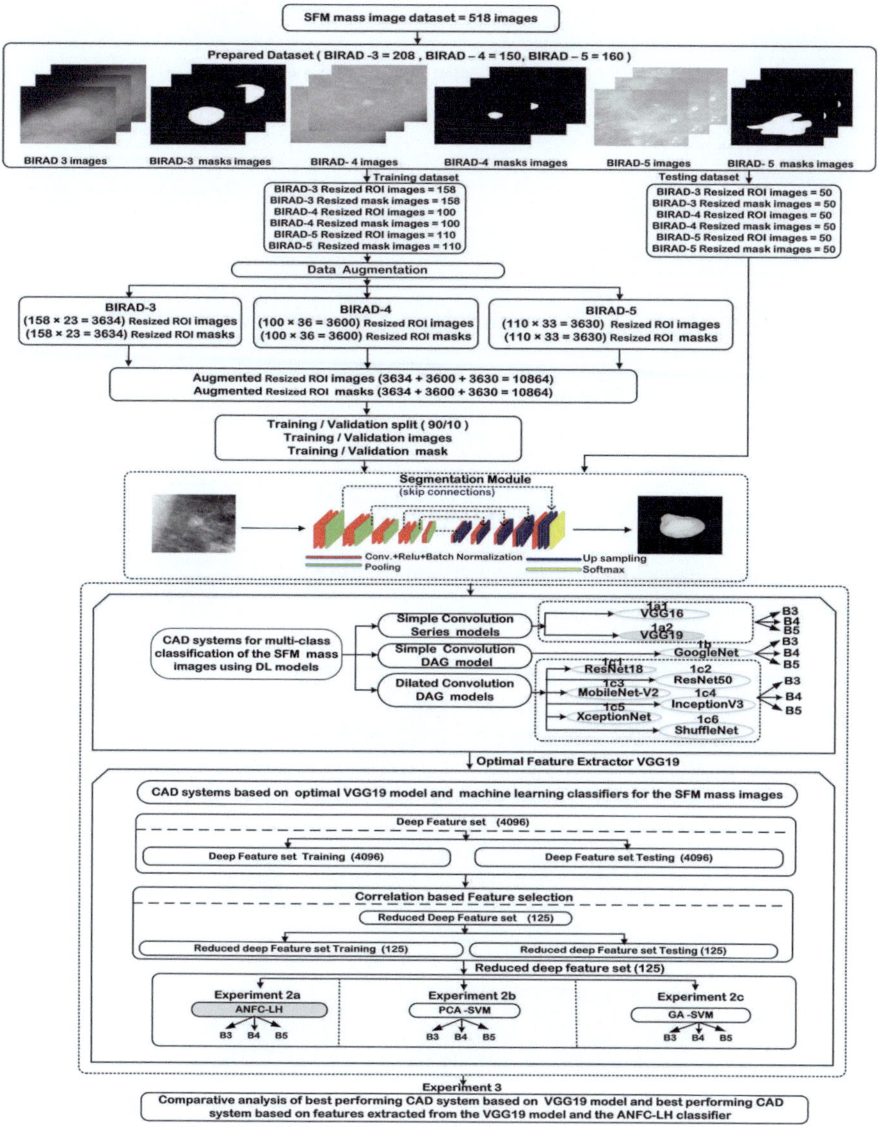

Fig. 5 Experimental methodology for multi-class classification. *Note B3*: BIRAD-3, *B4*: Suspicious malignant, *B5*: Highly malignant

3.2 *ROI Segmentation*

In present study, the ROI segmentation from SFM mass images has been carried out using DL-based ResNet50 DAG model [6, 48, 66] the ROI segmentation is shown in Fig. 7.

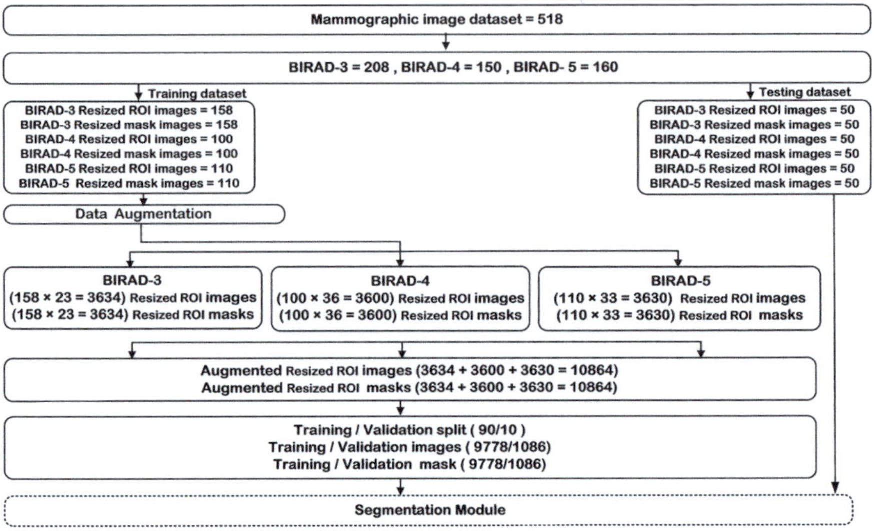

Fig. 6 The dataset description for BIRAD-3, BIRAD-4, BIRAD-5 class

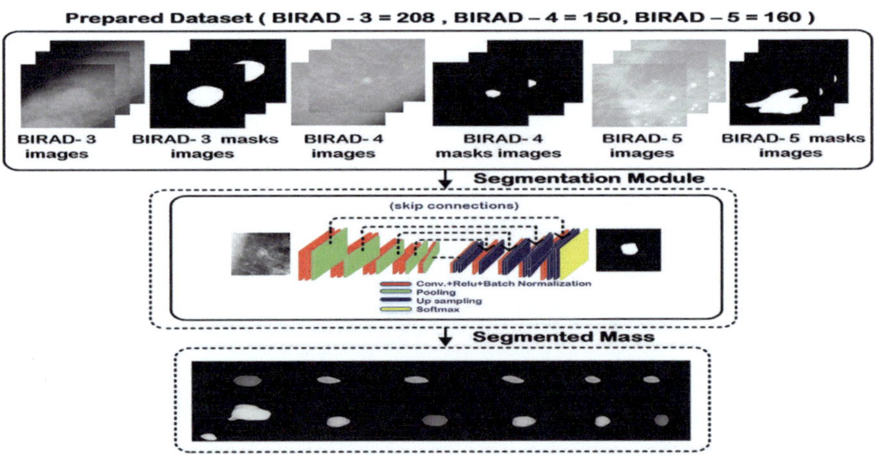

Fig. 7 ROI segmentation

Implementation Details: The present work inculpates, the DL-based models which have been trained for multi-class classification of the SFM mass images using a dataset of total 10,864 augmented training images. Present study has used 10-fold cross validation to train the DL models, i.e. the training /validation split is 90% / 10%. The fine tuning of the DL models was carried out using adam optimizer by varying learning rate values-0.01, 0.001 as well as 0.0001 with batch size-32 and epochs-30. The system used in the present work have GPU NVIDIA GeForce GTX 1070Ti, Intel i7 processor with 3.8 GHz having 32 GB RAM and MATLAB version

R2019b with DL tool-box has been used for implementing the DL based models. The ML based classifiers have been implemented using MATLAB 2015b software.

Performance Evaluation Metrics: The DL-based models are evaluated with overall class accuracy and individual class accuracy (B3), individual class accuracy (B4) and individual class accuracy (B5) metric [7, 17, 34, 50, 51]. The evaluation metrics evaluated in the present work are given in Eq. (1), (2), (3) and (4) and is as follows:

$$\text{OCA} = \frac{\text{CCI}}{\text{TTI}} \times 100 \tag{1}$$

$$\text{ICA (B3)} = \frac{\text{CCI(B3)}}{\text{TTI(B3)}} \times 100 \tag{2}$$

$$\text{ICA (B4)} = \frac{\text{CCI(B4)}}{\text{TTI(B4)}} \times 100 \tag{3}$$

$$\text{ICA (B5)} = \frac{\text{CCI(B5)}}{\text{TTI(B5)}} \times 100 \tag{4}$$

Note ICA = Individual class accuracy, *CCI (B* 3) = Correctly Classified Instances of B3 class, *CCI (B* 4) = Correctly Classified Instances of B4 class, *CCI (B* 5) = Correctly Classified Instances of B5 class, *TTI (B3)* = total testing instances of B3 class, *TTI (B4)* = total testing instances of B4 class, *TTI (B5)* = total testing instances of B5 class, *CCI* = Correctly Classified Instances, *TTI* = total testing instances, *B3*: BIRAD-3, *B4*: BIRAD-4, *B5*: BIRAD-5.

4 Experiments Details

The segmented mass images obtained from the segmentation model ResNet50 have been used for the various CAD systems [48] as motioned in the Table 4.

4.1 Experiment Results and Discussion

Experimentation results are discussed as follows:

(i) ***Experiment 1a–1c*: CADs for multi-class classification of SFM mass images with DL-models**

The overall classification accuracy and individual class accuracy for BIRAD-3, BIRAD-4 and BIRAD-5 classes, respectively for multi-class classification of DL-based CAD system is mentioned in Table 5.

Table 4 Exhaustive experiments conducted in the present study

	CAD systems for multi-class classification of SFM mass images with DL-models
Experiment 1a	Simple convolution series models
Experiment 1a1	VGG16 model
Experiment 1a2	VGG19 model
Experiment 1b	CAD system based on simple convolution DAG model using GoogleNet model
Experiment 1c	CAD systems based on dilated convolution DAG models
Experiment 1c1	ResNet18 model
Experiment 1c2	ResNet50 model
Experiment 1c3	MobileNet-V2 model
Experiment 1c4	Inceptionv3model
Experiment 1c5	XceptionNet model
Experiment 1c6	ShuffleNet model
Experiment 2a	CAD systems based on optimal VGG19 model and machine learning classifiers for the SFM mass images using ANFC-LH
Experiment 2b	PCA-SVM
Experiment 2c	GA-SVM
Experiment 3	Comparative analysis of best performing CAD system based on the VGG19 model and best performing CAD system based on features extracted from the VGG19 model and the ANFC-LH classifier

Concluding Remarks: Table 5 concludes that the DL-based CAD system with VGG19 Model yields highest classification accuracy of 84.6%, individual BIRAD-3 class accuracy of 98%, individual BIRAD-4 class accuracy of 80% and individual BIRAD-5 class accuracy of 76%.

(ii) ***Experiment 2a–2c*: CAD systems based on optimal VGG19 model and machine learning classifiers for SFM mass images**

The objective assessment of different DL-based models verified that VGG19 model is an optimal feature extractor with highest accuracy of 85% and is the deep feature-set extractor. For the detailed description of VGG19 model the readers are directed

Table 5 Objective assessment of DL models for multi-class classification of the SFM mass images

CNN model	Confusion matrix				Overall classification Accuracy (%)	Individual class accuracy for B3 (%)	Individual class accuracy for B4 (%)	Individual class accuracy for B5 (%)
VGG16 (*experiment 1a1*)		B3	B4	B5	81	96	76	72
	B3	48	2	0				
	B4	6	38	6				
	B5	9	5	36				
VGG19 (*experiment 1a2*)		B3	B4	B5	84.6	98	80	76
	B3	49	0	1				
	B4	8	40	2				
	B5	7	5	38				
GoogLeNet (*experiment 1b*)		B3	B4	B5				
	B3	46	1	3	80.6	92	70	80
	B4	7	35	8				
	B5	6	4	40				
ResNet18 (*experiment 1c1*)		B3	B4	B5	80.6	96	70	76
	B3	48	1	1				
	B4	7	35	8				
	B5	7	7	38				
ResNet50 (*experiment 1c2*)		B3	B4	B5	82	94	76	76
	B3	47	1	2				
	B4	6	38	6				
	B5	6	6	38				
MobileNet-v2 (*experiment 1c3*)		B3	B4	B5	78.6	92	68	74
	B3	46	7	8				
	B4	3	34	5				
	B5	1	9	37				
Inceptionv3 (*experiment 1c4*)		B3	B4	B5	72.6	86	76	56
	B3	43	4	3				
	B4	6	38	8				
	B5	10	12	28				
XceptionNet (*experiment 1c5*)		B3	B4	B5	70.6	84	68	60
	B3	42	6	2				
	B4	8	34	8				
	B5	11	9	30				
ShuffleNet (*experiment 1c6*)		B3	B4	B5	75.3	84	66	76
	B3	42	4	4				

(continued)

Table 5 (continued)

CNN model	Confusion matrix				Overall classification Accuracy (%)	Individual class accuracy for B3 (%)	Individual class accuracy for B4 (%)	Individual class accuracy for B5 (%)
	B4	12	33	5				
	B5	7	5	38				

Note B3: *B3*: BIRAD-3, *B4*: BIRAD-4, *B5*: BIRAD-5

to [48]. The extracted deep feature-set may have the redundant values [22, 33, 48]. These redundant values may affect the multi-class classification using the SFM mass images [10, 40]. The deep feature-set comprising of 4096 features is inputted to correlation based feature selection method to yield (reduced) deep feature-set having 125 selected features, which is further inputted to various ML-classifiers [48]. For the details of DL-based models and ML-based classifiers used in the present work, the readers are directed to [48].

ANFC-LH: Fuzzy logic was implemented by Zadeh in 1965[77, 78]. Every pixel is assigned a membership that shows to which class it belongs to. The concept of fuzzy classification allows defining every pixel in terms of its membership to all the other classes. Membership function can be used to define the percentage of the pixels that belongs to other classes. Fuzzy set theory is used in fuzzy networks. Expert knowledge is used to creating and modifying the membership values. Linguistic variables are very much useful in reasoning and the linguistic hedge incorporates the fuzzy rules [12, 13].

It basically combines the learning capability of the neural-network with the capability of knowledge representation of fuzzy logics to yield fuzzy neural networks [67]. Adaptive neural fuzzy inference system is prominent part of fuzzy neural networks, which is sugeno fuzzy model based. They address the issue of class overlapping resulting in reduced optimal feature vector.

PCA-SVM: PCA is a feature reduction technique. As all the features present may not be equally important and may have redundant feature values and classification model need not to be overloaded with redundant information. There is need to reduce the features without extracting out the useful information from the data. PCA is a popular technique for dimensionality reduction.

PCA does the conversion of correlated-feature variables into linear uncorrelated-features known as principal components [72]. The number of PCs will be equal to number of dimensionality of input data.

SVM is the popular machine learning algorithm, used to create a best line or the decision boundary that segregates the n-dimensional space in various classes to which various data points can be correctly put on. PCA-SVM combination provides the benefit of decreasing computational time with increased efficiency. PCA-SVM also reduces the over-fitting problem [38, 72].

GA-SVM: GA-SVM is an optimization method that uses iteration for finding the optimal solution [65]. Initially, for the genetic problem a random solution is given which is examined for its convergence toward optimal solution and it meets the termination condition, otherwise next generation is generated by evaluating the fitness for initial population. The reproduction phase will use crossover to generate the off-springs. The mutation will be used to generate new off-springs. The generated muted off-springs will be checked for termination condition and finally convergence for optimal solution is met with after several iterations [65]. The optimization towards the optimal solution is incorporated with SVM algorithm.

The objective evaluation metrics yielded by CADs with an optimal VGG19 model and ML-based classifiers have been reported in Table 6.

From Table 6, it is clear that CADs having VGG19 model with ANFC-LH classifier, reports the overall classification accuracy of 86%, individual class accuracy of 98% for BIRAD-3 class, individual class accuracy of 84% for BIRAD-4 class and individual class accuracy of 76% for BIRAD-5 class, respectively.

(iii) ***Experiment 3*****: Comparative analysis of best performing CAD system based on the VGG19 model and best performing CADs on features extracted from the VGG19 model and the ANFC-LH classifier.**

Comparative analysis of best performing VGG19 feature-extractor and VGG19 model with ANFC-LH classifier is mentioned in Table 7.

It can be concluded, the CADs with VGG19 model and ANFC-LH classifier gives highest overall classification accuracy of 86% for multi-class classification of SFM mass images with the individual class accuracy values 98, 84 and 76% for B3, B4 and B5 class, respectively.

Table 6 Objective evaluation of the CAD system using an optimal VGG19 model and Machine Learning based classifiers

CNN model	Confusion matrix				Overall classification accuracy (%)	Individual class accuracy for B3 (%)	Individual class accuracy for B4 (%)	Individual class accuracy for B5 (%)
VGG19 with ANFC-LH classifier (*experiment 2a*)		B3	B4	B5	86	98	84	76
	B3	49	1	0				
	B4	5	42	3				
	B5	7	5	38				
VGG19 with PCA-SVM classifier (*experiment 2b*)		B3	B4	B5	83.3	98	82	70
	B3	49	1	0				
	B4	7	41	2				
	B5	7	8	35				
VGG19 with GA-SVM classifier (*experiment 2c*)		B3	B4	B5	82	98	80	68
	B3	49	1	0				
	B4	7	40	3				
	B5	9	7	34				

Note B3: BIRAD-3, *B4*: BIRAD-4, *B5*: BIRAD-5

Table 7 Misclassified image analysis of CADs using VGG19 and CAD system on optimal VGG19 model with ANFC-LH classifier

CAD system	OCA (%)	Misclassified cases as per class			No. of misclassified cases
		B3	B4	B5	
VGG19	84.6	01/50	10/50	12/50	**Total = 23/150**
VGG19 with ANFC-LH classifier	86	01/50	08/50	12/50	**Total = 21/150**

Note B3: BIRAD 3, *B4:* BIRAD 4, *B5:* BIRAD 5, *OCA*: Overall classification accuracy

The misclassification image analysis of CADs using the best performing deep feature-extractor VGG19 and ANFC-LH is done by the experienced participating radiologist by observing the (*i*) background breast tissue density (*ii*) shape and (*iii*) margins properties exhibited by misclassified SFM mass images. It has been validated by participating expert radiologist that the images with irregular shape, spiculated margins lead to most of the misclassified image cases. Amongst the misclassified cases of the CADs using optimal feature-extractor VGG19 with ANFC-LH, it is validated by the participating radiologist that, out of all the misclassified images, 04 images were commonly misclassified which includes 02 images of B4 and 02 images of B5 class. It is pertinent to mention that these cases of suspicious abnormalities (i.e. B4 and B5 classes) have been misclassified as probably benign i.e. B3 class. These commonly misclassified cases are shown in Fig. 8.

It was observed that the desmoplastic reaction has caused few suspiciously malignant (B4) and highly malignant cases (B5) to be misclassified as probably benign cases. Probably because of the undergoing desmoplastic reaction due to which there is a growth of fibrous tissue along the tumor cells which adversely affects the background density of the tissue leading to masking of margin and shape characteristics exhibited by malignant tumors.

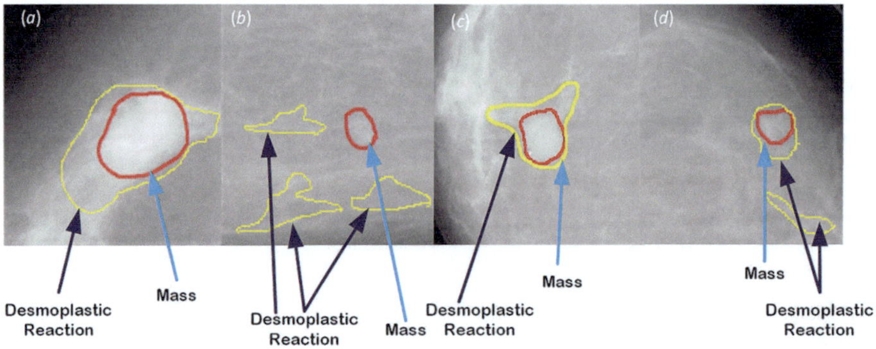

Fig. 8 Commonly misclassified cases **a** A_1820_1. LEFT_CC is a case of B4 misclassified as B3, **b** A_1121_1. LEFT_CC is a case of B5 misclassified as B3; **c** C_1171_1.LEFT_CC is a case of B5 misclassified as B3 **d** C_1640_1.RIGHT_CC is a case of B5 misclassified as B3

5 Conclusion

From the experimentation conducted in present work, it is concluded that deep features extracted by VGG19 Model and the ANFC-LH classifier gives the highest accuracy of 86% for multi-class classification of SFM mass images with ICA values of 98, 84 and 76% for B3, B4 and B5 classes, respectively. From the subjective analysis, it was observed that the desmoplastic reaction has caused few suspiciously malignant (B4) and highly malignant cases (B5) to be misclassified as probably benign (B3) cases. Probably, because of the undergoing desmoplastic reaction due to which there is a growth of fibrous tissue along the tumor cells which adversely affects the background density of the tissue leading to masking of margin and shape characteristics exhibited by malignant tumors.

Acknowledgements The authors feels immense pleasure to thank the expert participating Radiologist Dr. Hargurjeet Singh, M.D. Radio-diagnosis, Max Hospital, Bathinda, Punjab, India for providing detailed understanding of screen film mammographic images and for carrying out the misclassification image analysis of the present research work.

Author Contribution 1: Writing—original draft, Survey Protocol, Research Gaps, Critical Analysis, editing final draft., 2: Research Gaps, Critical Analysis, Writing—review & editing.,3: Research Gaps, Critical Analysis, Writing—review & editing.

Data Availability The dataset generated during and/or analyzed during the current study is available in https://publications.rwth-aachen.de/record/667223.

Conflict of Interest The authors declare that they have no conflict of interest.

References

1. American Cancer Society: Breast cancer (facts and figures). https://www.ajronline.org/doi/full/10.2214/AJR.17.18707
2. Anwar SM et al (2018) Medical image analysis using convolutional neural networks: a review. J Med Syst 42(226):1–20
3. Al-antari MA et al (2017) An automatic computer-aided diagnosis system for breast cancer in digital mammograms via deep belief network. J Med Biol Eng 38(3):443–456
4. Ansar W et al (2020) Breast cancer detection and localization using MobileNet based transfer learning for mammograms. In: International symposium on intelligent computing systems, vol 1187, pp 11–21
5. Aboutallib SS et al (2018) Deep learning to distinguish recalled but benign mammography images in breast cancer research. J Clin Cancer Res 24(23):5902–5909
6. Abdelhafiz D et al (2019) Residual deep learning for mass segmentation and classification in mammography. In: ACM international conference on bioinformatics, computational biology and health informatics, pp 475–484. https://doi.org/10.1145/3307339.3342157
7. Arias R et al (2019) Evaluation of learning approaches based on convolutional neural networks for mammogram classification. In: International conference on smart technologies, systems and applications, vol 1154, pp 273–287
8. Ballin AA et al (2016) A region based convolutional network for tumor detection and classification in breast mammography. In: International workshop on large-scale annotation of biomedical data and expert label synthesis, pp 197–205

9. Benzebouchi NE et al (2019) A computer-aided diagnosis system for breast cancer using deep convolutional neural network. Comput Intell Data Min 583–593. https://doi.org/10.1007/978-981-10-8055-5_52
10. Bektas B et al (2018) Classification of mammography images by machine learning techniques. In: IEEE international conference on computer science and engineering, vol 580–585
11. Carneiro G et al (2019) Automated analysis of unregistered multi-view mammograms with deep learning. IEEE Trans Med Imaging 36(11):2355–2365
12. Cetisli B (2010) Development of an adaptive neuro-fuzzy classifier using linguistic hedges: part 1. Expert Syst Appl 37:6093–6101
13. Cetisli B (2010) Development of an adaptive neuro-fuzzy classifier using linguistic hedges: part 2. Expert Syst Appl 37:6102–6108
14. Debelee TG et al (2018) Classification of mammograms using convolutional neural network based feature extraction. In: International conference on information and communication technology for development for Africa, vol 244, pp 89–98
15. Forsyth D et al (2002) Computer vision: a modern approach. Prentice Hall Professional Technical Reference
16. Dabbas J et al (2021) Multi-class classification of mammograms with hesitancy based Hanman transform classifier on pervasive information set texture features. Informatics in Medicine Unlocked 26:1–14
17. Ghazouni H et al (2021) Towards non-data-hungry and fully-automated diagnosis of breast cancer from mammographic images. Comput Biol Med 139:105011
18. Guan S et al (2017) Breast cancer detection using transfer learning in convolutional neural networks. In: IEEE applied imagery pattern recognition workshop, pp 1–8
19. Garcia AH et al (2018) Further advantages of data augmentation on convolutional neural networks. In: International conference on artificial neural networks, vol 11139, pp 95–103
20. Gau Y et al (2016) Deep learning for visual understanding: a review. Neurocomputing 187:27–48
21. Gananasekaran VS et al (2020) Deep learning algorithms for breast masses classification in mammograms. J Inst Eng Technol Image Process 14(12):2860–2868
22. Gardezi SJS et al (2017) Mammogram classification using Deep learning features. In: IEEE international conference on signal image processing applications, pp 485–488
23. Hang W et al (2017) GlimpseNet: attentional methods for full-image mammogram diagnosis. Comput Sci Med. https://api.semanticscholar.org/CorpusID:201648327
24. Hussain Z et al (2018) Differential data augmentation techniques for medical image classification tasks. In: Annual symposium proceedings archive, pp 979–984
25. Heath M et al (2000) Current status of digital database for screening mammography. In: International conference on digital mammography, pp 212–218
26. Hu Z et al (2018) Deep learning for image-based cancer detection and diagnosis-a survey. Pattern Recogn 83:134–149
27. Howard AG et al (2017) Mobilenets: efficient convolutional neural networks for mobile vision applications. Int J Comput Vis Pattern Recognit 1–9. https://doi.org/10.48550/arXiv.1704.04861
28. Jadoon MM et al. "Three-class mammogram classification based on descriptive CNN features", International Journal of Biomedical Research, PP 1–12, 2017.
29. Jiao Z et al (2016) A deep feature based framework for breast masses classification. J Neurocomputing 197:221–231
30. Kriti et al (2020) Deep feature extraction and classification of breast ultrasound images. Multimed Tools Appl 79:27257–27292
31. Kavitha T et al (2021) Deep learning based capsule neural networks for breast cancer diagnosis using Mammogram images. Interdiscip Sci: Comput Life Sci 14:113–129. https://doi.org/10.1007/s12539-021-00467-y
32. Kriti et al (2015) Breast density classification using Laws' mask texture features. Int J Biomed Eng Technol 19(3):279–302
33. Lecon Y et al (2015) Deep learning. Nature 521:436–444

34. Levy D et al (2016) Breast mass classification from mammograms using deep convolutional neural network. In: International conference on neural information processing systems, pp 1–6
35. Li H et al (2020) Classification of breast mass in two view mammograms via deep learning. Inst Eng Technol Image Process 15:15454–15467
36. Lotter W et al (2017) A multi-scale CNN and curriculum learning strategy for mammogram classification deep learning in medical image analysis and multimodal learning for clinical decision support. J Springer 10553:169–177
37. Li H et al (2012) Computerized Analysis of Mammographic parenchymal patterns on a large clinical dataset on FFDM robustness study with two high risk datasets. J Digit Imaging 25(5):591–598
38. Mustaqeem M et al (2021) Principal components based support vector machine: a hybrid technique for software defect detection. Clust Comput 24:2581–2595
39. Sandler M et al (2018) Mobilenetv2: inverted residuals and linear bottlenecks. In: IEEE conference on computer vision and pattern recognition, pp 4510–4520
40. Michalak K et al (2006) Correlation-based feature selection strategy in classification problems. Int J Appl Math Comput Sci 16:503–511
41. Malberry SJ et al (2021) Automated breast mass classification system using DL and ensemble learning. IEEE Access 9:55312–55328
42. Mustra M et al (2012) Breast density classification using multiple feature selection. Automatika 53(4):362–372
43. Nagaraj M et al (2019) Classification of mammograms using attention learning or localization of malignancy. Int J Eng Adv Technol 8(5S):84–90
44. Rampun A et al (2018) Breast mass classification in mammograms using ensemble convolutional neural networks. In: IEEE international conference on e-health networking, application & services, pp 1–6
45. Rani J et al (2023) Mammographic mass classification using DL based ROI segmentation and ML based classification. In: International conference on device intelligence, computing and communication technologies, pp 302–306
46. Ray KM et al (2018) Screening mammography in women 40–49 years old: current evidence. Am J Roentgenol 210(2):264–270
47. Ribli D et al (2018) Detecting and classifying lesions in mammograms with deep learning. Sci Rep 8(4165):1–7. https://doi.org/10.1038/s41598-018-22437-z
48. Rani J et al (2023) Hybrid computer aided diagnostic system design for screen film mammograns using DL-based feature extraction and ML-based classifiers. Expert Syst e13309:1–29. https://doi.org/10.1111/exsy.13309
49. Song R et al (2020) Mammographic classification based on XGBoost and DCNN with multi features. IEEE Access 8:75011–75021
50. Saranyaraj D et al (2020) Detecting and classifying lesions in mammograms with deep learning. Multimed Tools Appl 79:11013–11038
51. Szegedy C et al (2015) Going deeper with convolutions. In: IEEE conference on computer vision and pattern recognition, pp 1–9
52. Shen D et al (2017) Deep learning in medical image analysis. Annu Rev Biomed Eng 19(1):221–248
53. Simonyan K et al (2015) Very deep convolutional networks for large scale Image recognition, pp 1–14
54. Saeed DM et al (2021) Early breast cancer diagnostics based on hierarchical machine learning classification for mammography images. Cogent Eng 8(1):1–19
55. Samala RK et al (2017) Multi-task transfer learning deep convolutional neural network application to computer-aided diagnosis of breast cancer on mammograms. J Phys Med Biol 62(23):8894–8908
56. Siegel RL et al (2020) Cancer statistics. CA Cancer J Clin 70(1):7–30
57. Shen L et al (2017) End-to-end training for whole image breast cancer screening using an all convolutional design. Sci Reports 9:1–7. arXiv:1708.09427

58. Sun L et al (2019) Multi view convolutional neural network for mammographic image classification. IEEE Access 7:126273–128282
59. Shamy S et al (2019) A research on detection and classification of breast cancer using K means GMM & CNN algorithms. Int J Eng Adv Technol 8(65):501–505
60. Sharma S et al (2015) Computer aided diagnosis of malignant mammograms using Zernike moments and SVM. J Digit Imaging 28:77–90
61. Szegedy C et al (2016) Rethinking the inception architecture for computer vision. In: IEEE conference on computer vision and pattern recognition, vol 1, pp 2818–2826
62. Shruthishree SH et al (2021) AlexResNet+: a deep hybrid featured machine learning model for breast cancer tissue classification. Turk J Comput Math Educ 12(6):2420–2438
63. Tariq M et al (2020) Medical image based breast cancer diagnosis: state of the art and future directions. Expert Syst Appl 167(114095):1–71
64. Tang CM et al (2019) Five classifications of mammography images based on deep cooperation convolutional neural network. Am Sci Res J Eng Technol Sci 57(1):10–21
65. Tao Z et al (2019) GA-SVM based feature selection and parameter optimization of hospitalization expense modeling. Appl Soft Comput J 75:323–332
66. Thuy TL et al (2019) Multitask classification and segmentation for cancer diagnosis in mammography. In: International conference on medical imaging with deep learning, pp 1–4
67. Ubeyli ED (2009) Adaptive neuro-fuzzy inference systems for automatic detection of breast cancer. J Med Syst 33:353–358
68. Wu J et al (2021) DeepMiner: discovering interpretable representations for mammogram classification and explanation. Harv Data Sci Rev 3(4):1–13
69. Williams LJ et al (2010) Principal component analysis. WIREs Comput Stat 8:433–459
70. Wang G (2016) A perspective on deep imaging. IEEE Access 4:8914–8924
71. Wang J et al (2017) The effectiveness of data augmentation in image classification using deep learning, pp 1–8. arXiv:1712.04621v1
72. Yang L et al (2019) Feature extraction by PCA and diagnosis of breast tumors using SVM with DE-based parameter tuning. Int J Mach Learn Cybern 10(3):591–601
73. Yang S (2017) Automated breast cancer diagnosis using deep learning and region of interest detection (BC-DROID). In: International conference on bioinformatics computational biology and health informatics, pp 536–543
74. Zhang C et al (2020) New convolutional neural network model for screening and diagnosis of mammograms. J PLOS ONE 15(8):1–20
75. Zhang X et al (2018) Shufflenet: an extremely efficient convolutional neural network for mobile devices. In: IEEE conference on computer vision and pattern recognition, pp 6848–6856
76. Zhou H et al (2017) Mammogram classification using convolutional neural networks. In: International conference on robotics, automation and sciences, pp 1–8
77. Zadeh LA (1975) The concept of a linguistic variable and its application to approximate reasoning-II. Inf Sci 8(4):301–357
78. Zadeh LA (1975) The concept of a linguistic variable and its application to approximate reasoning-III. Inf Sci 9(1):43–80

Enhancing Video Surveillance: Synergizing Time-Distributor Wrapper and Attention Mechanisms for Superior Human Action Recognition

Khawla Ben Salah, Mohamed Othmani, Jihen Fourati, and Monji Kherallah

Abstract This paper presents a novel approach for human action recognition in videos by focusing on the extraction of spatio-temporal features. Recognizing actions in videos is of paramount importance for various applications, including video surveillance, sports analysis, and human-computer interaction. To address this challenge, we proposed an hybrid network that integrates a time-distributed wrapper and attention-based mechanisms. The incorporation of a time-distributed wrapper enables the model to effectively extract the temporal dynamics of actions by extending the capabilities of Convolutional Neural Networks (CNNs) to process sequences of frames. Moreover, the integration of attention-based mechanisms allows the network to selectively weigh and emphasize relevant spatio-temporal features, further enhancing its discriminative power. The proposed network outperforms the state-of-the-art methods in human action recognition, achieving an impressive average accuracy of 99.3% and 99.49% on the UCF101 and UCF11 datasets respectively.

Keywords Human action recognition · Video surveillance · UCF11 · Attention mechanism

---

K. B. Salah (✉) · M. Othmani · J. Fourati · M. Kherallah
Computer Sciences, ATES: Advanced Technologies onEnvironment and Smart City, National Engineering School, Sfax, Tunisia
e-mail: khawla.bensalah@fsgf.u-gafsa.tn

M. Othmani
e-mail: mohamed.othmani@FSGF.u-gafsa.tn

J. Fourati
e-mail: jihen.fourati@enis.u-sfax.tn

M. Kherallah
e-mail: monji.kherallah@fss.usf.tn

Computer Sciences, ATES: Advanced Technologies onEnvironment and Smart City, University of Gafsa, Gafsa, Tunisia

National engineering school of Sfax, University of Sfax, Sfax, Tunisia

Physics, ATES: Advanced Technologies on Environment andSmart City, University of Sfax, Sfax, Tunisia

S. Chand Kumain et al. (eds.), *AI and ML Techniques in Image Processing and Object Detection*, https://doi.org/10.1007/978-981-96-7445-9_11

1 Introduction

Video-based Human Action Recognition (HAR) [1] has gained significant attention due to its wide range of potential applications. For example in video surveillance and security applications by accurately recognizing and classifying human actions. It is crucial to these applications to identify suspicious or abnormal activities, aiding in the detection of potential threats or criminal behavior in public spaces, airports, or critical infrastructure. Also, (HAR) finds relevance in fields like sports analysis [2], where it assists in tracking and analyzing athletes' movements and actions by automatically recognizing and quantifying actions such as running, jumping, or throwing. This option provides valuable insights into athletes' performance, enabling performance evaluation, training optimization, and even referee assistance in certain sports. Furthermore, human action recognition has implications in healthcare [3] and assisted living. (HAR) introduces numerous challenges such as camera movements, occlusions, complex backgrounds, and variations in illumination. Spatial and temporal information are critical in accurately recognizing different human actions in videos. In the past decade, many methods relied on handcrafted feature engineering to represent the spatial attributes of dynamic motion in video actions. However, these handcrafted approaches are often limited to specific databases and struggle to handle diverse motion styles and complex backgrounds. To address this, there has been a shift towards upgrading representative motion features and conventional methods from 2D to 3D. This enables the extraction of accurate information by moving from spatial features into 3D spatiotemporal features, allowing for the simultaneous analysis of dynamic information across a sequence of frames. In this paper, we proposed a novel Hybrid network with a time-distributor wrapper and Attention-Based Mechanism. This approach aims to extract spatiotemporal features and focus on long-term sequences for accurate action recognition in video frames. In our study, we employed the Long-term Recurrent Convolutional Networks (LRCN) model [4] with attention mechanism, which specifically accentuate efficient features within the video's frames sequence to recognize actions in the video effectively. The time-distributor wrapper helps extract the temporal evolution of actions by distributing information across different time steps, while the attention-based mechanisms focus on relevant features within the video frames sequence. The attention mechanisms enable the model to selectively attend to informative regions or frames, improving its understanding of the spatiotemporal dynamics in sequential data, and it helps the model focus on relevant information for accurate action recognition. The paper's structure is systemized as follows: the related works are elucidated in Sect. 2; The proposed methodology is depicted in Sect. 3; The results and the discussion are demonstrated in Sect. 4; finally, the conclusion is represented in Sect. 5.

2 Related Work

Deep learning is commonly used in video-based action and behavior detection to extract high-level discriminative characteristics. Dai et al. [5] introduced a two-stream LSTM architecture for action recognition: a spatial stream and a temporal stream. The spatial stream processes the visual appearance of individual frames, while the temporal stream extract the motion information across frames. Meng et al. [6] proposed an approach for action recognition that addresses the limitations of existing convolutional neural network (CNN) models. The authors combine a quaternion spatial-temporal CNN (QST-CNN) with a Long Short-Term Memory (LSTM) network. The authors in [7] proposed a framework that combines convolutional neural networks (CNNs) to learn spatial features and maps their temporal relationships using Long-Short-Term-Memory (LSTM) networks. Similarly, In their study [8], the authors also focus on the fusion of different features for human action recognition. They proposed six fusion models inspired by early fusion, late fusion, and intermediate fusion schemes. The first two models in their approach employ the early fusion technique, where they combine multiple modalities or features at an early stage. The third and fourth models employ intermediate fusion techniques, involving the combination of features or decision scores at an intermediate level. Specifically, the fourth model incorporates a kernel-based fusion scheme, utilizing the kernel basis of classifiers such as Support Vector Machine (SVM). Gharaee et al. [9] proposed an approach that combines the benefits of Self-Organizing Maps (SOMs), supervised neural networks, and attention mechanisms to create an action recognition system. Their approach specifically focuses on leveraging joint movement dynamics in action recognition. They also introduce a custom supervised neural network that learns to classify actions effectively. Pan et al. [10] proposed an approach for recognizing human actions in basketball scenarios. They employed a motion region selection method based on constructing a large affinity graph to identify relevant regions of motion. In order to extract features from these motion blocks, they employed Gaussian Mixture Models (GMM). Additionally, the authors employed a variation modeling technique to select key frames that determine the variations between adjacent frames. They represented the posture descriptor of basketball actions using the gradient histogram, which allowed them to calculate a shape descriptor for basketball action recognition, they linearly combined the motion and posture descriptors and the K-Nearest Neighbors (KNN) algorithm. Muhammed et al. [11] proposed a novel approach that combines a bi-directional long short-term memory (BiLSTM) based attention mechanism with a dilated convolutional neural network (DCNN). This integrated model selectively focuses on relevant features within input frames to accurately classify various human actions in videos. The DCNN layers are employed to extract discriminative features using residual blocks, which help retain more information compared to shallow layers. Also authors in [12] involves Dense Semantics-Assisted CNN architecture utilizing dense semantic segmentation masks for improved human action recognition. Kamel et al. [14] proposed two types of data sequences used as input, namely joint posture sequence and depth map sequence. After transforming them into descriptors, the descriptor used for body posture is a proposed moving joints descriptor (MJD)and that used for depth map is DMI. Then, the input preprocessing is done, and three CNN models are trained with three different channels (Ch1, Ch2, and Ch3) and they are tested with different inputs. In the three CNN channels, one is trained for depth map images, another is trained with joint postures, and another is trained with both joint postures and depth map images. Using the score fusion operation, all the outputs are fused, and the final action is classified. Jaouedi et al. [15] have mainly focused on analyzing human behavior from recorded data from a camera or any other electronic source, and they have also paid attention to background actions such as fast walking and sudden movements. This model has been mainly designed to predict human behavior through the analysis of their movements. In this study, they have explained the recognition of human actions using the k-nearest neighbors (KNN) approach. In this study, they have used the GMM, or Gaussian mixture model, which is generally used for data analysis. GMM mainly focuses on the areas where the current state of the pixel changes from the previous state in a sequence of image collections. The proposed algorithm runs on each frame image converted into binary images for better performance. For this, they have declared 0 for black corresponding to their background and 1 for white corresponding to the background. The Kalman filter method is used for tracking human movements. And these filters are used frequently in two phases, namely prediction and correction. In the prediction phase, the current state is calculated using the information of the previous state. The main objective of this study is to obtain an efficient output. Finally, the classification is performed using the KNN method and has achieved a rate of 71.1%. Xio et al. [16] used a deep neural network model that uses an autoencoder, PRNN, or pattern recognition neural network to predict actions performed by humans. They used two approaches: a learning system and an action recognition stage. In the learning system, they created a binary frame for each image by drawing the contours of the human body and then joining all the frames. They used these frames to train their model. In another approach, they used an autoencoder to train the model to predict action features. After these two approaches, they trained the PRNN model using an unsupervised learning technique. Finally, they merged the autoencoder followed by the PRNN model, called APRNN. To evaluate the performance of APRNN, they used Weizmann motion data comprising 93 action clips recorded with 10 motion semantics. To improve the performance, they used fine tuning. Ji et al. [17] have mainly focused on human action analysis in robotic platforms. They have considered the different stages of human action recognition and prediction. In this paper, they have divided the field of human action recognition into three main categories: hand gesture-based human-robot interactions, body action-based human-robot interactions, and multimodal fusion. They have discussed the different platforms and datasets commonly used in the field of human-robot interaction. They have also addressed the different challenges and opportunities in the field of action analysis for human recognition. They have concluded that, in the future, data should be constructed to solve the storage problems related to data. Wang et al. [18] proposed a total of ten Kinect-based algorithms used on six datasets. These algorithms are aimed at multi-angle and multi-subject detection. The algorithms used are HON4D, HDG, LARP-SO, HOPC, SCK+DCK, P-LSTM, HPM+TM, clips+CNN+MTLN, indRNN, and ST-GCN. A 3D action analysis was also performed to compare the results of action recognition across objects and across view angles. It was concluded that depth-based action recognition techniques are better for recognizing objects with more details. They performed an extensive evaluation of HDG representation with different variants of descriptor types. They also introduced four variants of the P-LSTM framework.

3 Proposed Methodology

This section outlines the methodological framework adopted to conduct the study, with the primary aim of ensuring comprehensive coverage and rigorous evaluation of relevant research within its scope.

3.1 Preprocessing Data

In order to prepare the tested dataset for the training and the testing of the model, we performed various preprocessing steps: including resizing frames, specifying the sequence length, and normalizing the pixel values. Firstly, each video file from the dataset was resized to a fixed height and width. In our case, we have defined the dimensions as 64 pixels for both the height and width. Additionally, we considered twenty frames that will be fed to the model per sequence. Moreover, as part of the preprocessing, we will normalize the pixel values within the range of 0 to 1. This normalization technique involves dividing each pixel value by 255 and by performing these steps, we aimed to speed up the network's convergence during the subsequent training phase.

3.2 Proposed Hybrid Model Architecture

The Long-term Recurrent Convolutional Network (LRCN) model with attention is the proposed architecture for video action recognition task. It combines the power of Convolutional Neural Networks (CNNs) for spatial feature extraction and Recurrent Neural Networks (RNNs) for the extraction of temporal dependencies. The attention mechanism boosts the model's ability to focus on relevant video frames during processing. The model takes as input a sequence of video frames, where the input sequence is represented as a 4D tensor corresponding to the desired height and width to which the video frames are resized. The third dimension corresponds the RGB channels of the frame and the final dimension represents the chosen number of frames. The model architecture as displayed in Fig. 1a and b can be divided into

Fig. 1 Visual representations of UCF101 dataset samples

three main parts: the convolutional backbone, the (LSTM) layer, and the attention mechanism.

3.3 The Convolutional Backbone

The input layer of the model takes the input tensor with dimensions detailed in the Preprocessing data section then a series of 2D convolutional layers were applied to each frame in the input sequence using the Time Distributed Wrapper. Each 2D convolution layer is followed by an activation function (ReLU). The mathematical equation of the 2D convolution layer is described as follows:

$$\mathrm{R}[i, j] = \sum_{m=0}^{M-1} \sum_{n=0}^{N-1} \mathrm{k}[m, n] \times \mathrm{I}[i + m, j + n] \tag{1}$$

where R[i, j] represents the value of the output feature map at position (i, j), K is the convolutional kernel (also named the filter or the mask), and I represents the value at position (i+m, j+n) in the input feature map, with M and N representing the height and width of the kernel, respectively. The TimeDistributed wrapper is especially useful when working with recurrent neural networks (RNNs) or convolutional neural networks (CNNs) that process sequential data. In our case the input data consists of sequences, where each element in the sequence represents a different time step since the trained sequences are frames.

The TimeDistributed wrapper allows us to apply a layer or a set of layers to each frame in the sequence of frames independently. By doing so, the layers can learn patterns and representations specific to each time step.

The TimeDistributed layer handles iteration over the elements of the sequence and applies the wrapped layers to each element individually. This way, the layers receive inputs of shape at each time step and produce outputs of the same shape. For instance, it can be used to apply the convolutional layers to each frame separately, producing a sequence of feature maps. It is particularly important for the relationship or dependency between the elements of the sequence as it allows the model to extract spatial information across different time steps and maintain the temporal structure of the data. The mathematical equation for the TimeDistributed wrapper is represented as follows:

$$\text{TimeDistributed}(f)(\mathbf{X})_t = f(\mathbf{X}_t) \tag{2}$$

$\text{TimeDistributed}(f)(\mathbf{X})_t$ represents the output of the TimeDistributed wrapper applied to the input sequence $\mathbf{X}$ at time step t. $f(\mathbf{X}_t)$ represents the application of the function f to the input $\mathbf{X}_t$ at time step t.

In order to reduce computing time, a common practice is to include a max-pooling layer immediately after each 2D Convolutional layer. This assures down-sampling the feature maps and extraction of the most relevant information. After the wrapped max-pooling layer, the output is then passed through a dropout layer defined as (0.5). The purpose of the dropout layer is to randomly drop a certain portion of the activations during training, which helps prevent overfitting. This wrapped sequence of a 2D Convolutional layer followed by a max-pooling layer and dropout is repeated three more times to extract hierarchical features from the input data. In our four 2D Convolutional layers, we used 3×3 filters. The first, second, third, and fourth groups of layers were adopted with 16, 32, 64, and 64 filters, respectively. Additionally, max pooling with 4×4 filters was applied.

3.4 The Long Short Term Memory (LSTM) Layer

An RNN, or Recurrent Neural Network, has gained significant popularity in recent times due to its ability to effectively incorporate past frame information from a video sequence into the current frame, resulting in improved action recognition. Unlike traditional CNN models like AlexNet and VGG, which are primarily designed for image classification tasks, RNNs are specifically tailored for processing series or continuous data. By considering the temporal information inherent in a sequence, RNNs overcome the limitations of CNNs in capturing sequential patterns. However, RNNs are prone to encountering the "vanishing gradient" issue, where gradients decrease exponentially during training phase, making it difficult for the network to learn long-term dependencies. To address this challenge, a specialized type of RNN called Long Short-Term Memory (LSTM) is often employed. LSTMs are equipped with memory cells and gates that regulate the flow of information, allowing them to determine and retain relevant long-term dependencies in the data.

Before feeding the output tensor resulted of the previous convolutional backbone into the LSTM layer, a wrapped Flatten layer is applied to reshape the tensor. It flattens each frame of the feature maps into a 1D vector. The representation of the wrapped flatten layer is defined as follows:

$$\text{TimeDistributed(Flatten)}(X) = [F_{\text{flat}}(1), F_{\text{flat}}(2), \ldots, F_{\text{flat}}(t)] \tag{3}$$

where $F_{\text{flat}}(t)$ represents the flattened vector of the t-th frame F.

The next step is to pass the flattened tensor through an LSTM layer with 32 units, setting the return sequences parameter to True. This configuration allows the LSTM layer to retain and output the sequence of hidden states for each time step. The LSTM layer performs computations on the flattened tensor, utilizing its memory cells and gates to extract long-term dependencies and temporal patterns in the data. Each hidden state at a time step encodes relevant information from previous time steps, allowing the model to extract sequential dependencies and context within the input sequence. The equations for the mechanism of an LSTM layer: the forget gate, the input gate, the cell state, the updated cell state, the output gate, the hidden state are displayed in Eqs. (4) to (9) respectively:

$$f_t = \sigma(W_f \cdot [h_{t-1}, x_t] + b_f) \tag{4}$$

$$i_t = \sigma(W_i \cdot [h_{t-1}, x_t] + b_i) \tag{5}$$

$$\tilde{C}_t = \tanh(W_c \cdot [h_{t-1}, x_t] + b_c) \tag{6}$$

$$C_t = f_t \odot C_{t-1} + i_t \odot \tilde{C}_t \tag{7}$$

$$o_t = \sigma(W_o \cdot [h_{t-1}, x_t] + b_o) \tag{8}$$

$$h_t = o_t \odot \tanh(C_t) \tag{9}$$

3.5 *Attention Mechanism*

The attention mechanism or more precisely the scaled product attention mechanism is introduced to enhance the model's ability to focus on attentive spatio-temporal features of the input sequence from the LSTM outputs. The process starts by applying a Time-Distributed Dense layer that outputs a single value using the tanh activation function to the LSTM outputs. This transformation formulates a non-linear mapping to the outputs and helps capture complex relationships within the sequence. The resulting tensor from the attention mechanism is then flattened to simplify the subsequent computations. This flattening operation reshapes the tensor into a 2D representation while preserving the sequence information. Then, the softmax activation function is applied to the flattened tensor. This ensures the normalization ot the values across the time steps, producing attention weights that shows the relative importance of each element in the sequence. In order to address the compatibility issue with the LSTM outputs, the attention weights are repeated and reshaped using RepeatVector and Permute steps. These operations guarantees that each element of the LSTM outputs has its corresponding attention weight. The attention weights are then applied element-wise to the LSTM outputs using the Multiply() operation. This operation amplifies the LSTM outputs that have higher attention weights, effectively emphasizing the most relevant parts of the sequence. Finally, the sent representation tensor is obtained by an aggregation with the attention-weighted LSTM outputs along the time step axis.

In order to obtain the attention weights A= $\begin{bmatrix} a_1 \; a_2 \; \vdots \; a_n \end{bmatrix}$ using the scaled product attention mechanism, the following steps are performed:

1. Computing the similarity scores between the query vector **Q** and each key vector $\mathbf{k}_i$ using dot product: Q = $\begin{bmatrix} q_1 \; q_2 \; \vdots \; q_n \end{bmatrix}$, K = $\begin{bmatrix} k_1 \; k_2 \; \vdots \; k_n \end{bmatrix}$, V = $\begin{bmatrix} v_1 \; v_2 \; \vdots \; v_n \end{bmatrix}$ $\mathbf{s} = [\mathbf{Q} \cdot \mathbf{k}_1, \mathbf{Q} \cdot \mathbf{k}_2, \ldots, \mathbf{Q} \cdot \mathbf{k}_n]$.

2. Scaling the similarity scores by dividing by the square root of the dimension of the query vectors: $\mathbf{s}' = \frac{\mathbf{s}}{\sqrt{d_q}}$.

3. Applying the softmax function to obtain the attention weights: The corresponding attention weights matrix can be represented as:

$$QK^T = \begin{bmatrix} q_{1k_1}^T & q_{1k_2}^T & \cdots & q_{1k_n}^T \\ q_{2k_1}^T & q_{2k_2}^T & \cdots & q_{2k_n}^T \\ \vdots & \vdots & \ddots & \vdots \\ q_{nk_1}^T & q_{nk_2}^T & \cdots & q_{nk_n}^T \end{bmatrix}$$

$\mathbf{A} = \text{softmax}\left(\frac{\mathbf{QK}^T}{\sqrt{d_k}}\right)\mathbf{V}$

where Query Vectors: $\mathbf{Q} \in \mathbb{R}^{d_q \times n}$, Key Vectors: $\mathbf{K} \in \mathbb{R}^{d_k \times n}$, Value Vectors: $\mathbf{V} \in \mathbb{R}^{d_v \times n}$, and the Attention Weights: $\mathbf{A} \in \mathbb{R}^{n \times m}$.

4 Experiments

4.1 Dataset

We evaluate our model on two dataset: UCF101 and UCF11 [13]. The UCF-101 dataset is a popular benchmark dataset in the field of action recognition and computer vision. It was introduced by the University of Central Florida (UCF) and contains 101 action categories, making it a valuable resource for training and evaluating action recognition algorithms. The UCF11 dataset poses significant challenges for

Fig. 2 The proposed hybrid network for human action recognition

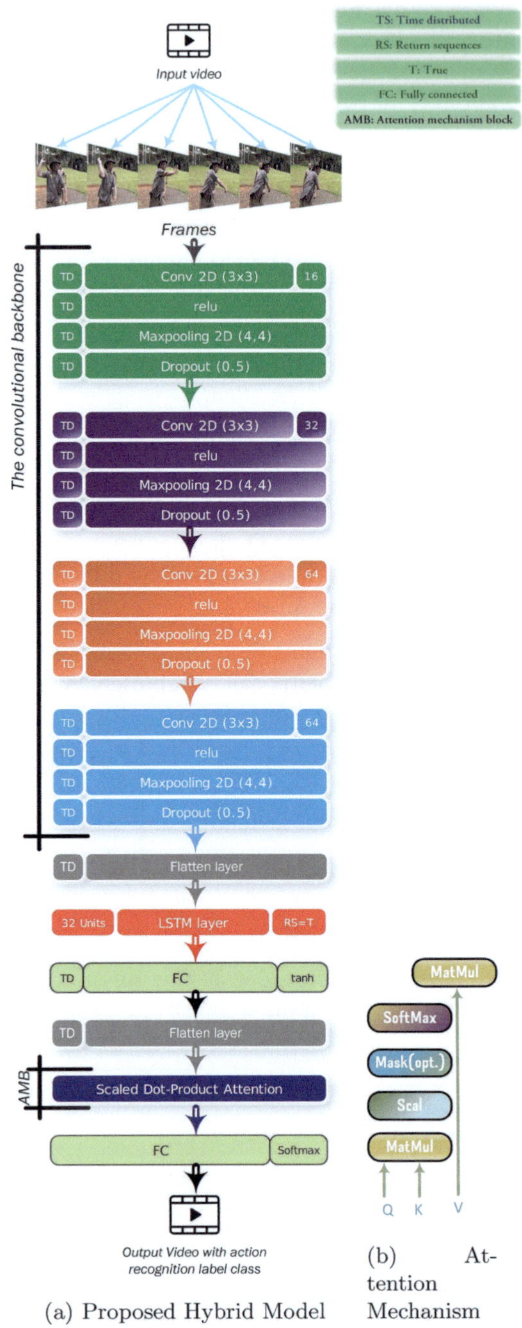

(a) Proposed Hybrid Model

(b) Attention Mechanism

(a) Basket Ball: Predicted action

(b) Fencing: Predicted action

(c) Walking with dog: Predicted action

(d) Taichi: Predicted action

(e) Swing: Predicted action

(f) Diving: Predicted action

Fig. 3 The proposed results: human action recognition from UCf-101 dataset

video-based action recognition due to various factors such as illumination variations, cluttered backgrounds, and camera movements. With a total of 1600 videos, the dataset comprises eleven action categories including shooting, jumping, riding, swimming, and more. The videos in the dataset are captured at a frame rate of 30 frames per second (fps). The presence of these challenges and the diversity of action categories make UCF11 a demanding benchmark for evaluating the performance of action recognition methods (Fig. 3).

4.2 Implementation Details

4.3 Quantitative Analysis

To assess the performance of the proposed framework, we conducted a comparative analysis, evaluating its performance comparing to other state-of-the-art methods. The Table 2 displays the performance evaluation of different human action recognition methods, including the proposed approach. The average accuracy values achieved by each method are listed, and the comparison is based on the UCF11 benchmark dataset. Looking at the previous works, Dai et al. introduced a two-stream LSTM architecture that leverages both spatial and temporal cues for action recognition. Their approach achieved an accuracy of 96.90%. Meng et al. addressed the limitations of CNN models by combining a quaternion spatial-temporal CNN (QST-CNN) with an LSTM network, resulting in a QST-CNN-LSTM architecture with an accuracy of 89.70%. Gammulle et al. proposed a fusion framework combining CNNs

Table 1 Parameters used in the proposed hybrid (HAR) network

Layer's name	Feature map dimensions	Kernel size
TimeDistributed (Conv2D) 1	(20, 64, 64, 16)	(3, 3)
TimeDistributed (MaxPooling2D) 1	(20, 16, 16, 16)	(4, 4)
TimeDistributed (Conv2D) 2	(20, 16, 16, 32)	(3, 3)
TimeDistributed (MaxPooling2D) 2	(20, 4, 4, 32)	(4, 4)
TimeDistributed (Conv2D) 3	(20, 4, 4, 64)	(3, 3)
TimeDistributed (MaxPooling2D) 3	(20, 2, 2, 64)	(2, 2)
TimeDistributed (Conv2D) 4	(20, 2, 2, 64)	(3, 3)
TimeDistributed (MaxPooling2D) 4	(20, 1, 1, 64)	(2, 2)
TimeDistributed (Flatten)	(20, 64)	–
LSTM	(20, 64)	
TimeDistributed (Dense)	(20, 1)	–
Flatten	(20)	–
Dense (Softmax)	101,11	–

and LSTM networks to capture both spatial and temporal features. Their CNN-LSTM model achieved an accuracy of 89.20%. Patel et al. focused on fusion models for action recognition, incorporating early, intermediate, and late fusion techniques. They achieved an accuracy of 89.43% using their fusion approach. Gharaee et al. utilized Self-Organizing Maps (SOMs), supervised neural networks, and attention mechanisms to effectively categorize actions. Their approach achieved an accuracy of 89.50%. Pan et al. developed a method for basketball action recognition, employing motion region selection, GMM-based feature calculation, and variation modeling. Their approach achieved an accuracy of 89.24%. In comparison, Muhammed et al. proposed the BiLSTM DCNN approach with an attention mechanism, achieving an accuracy of 98.30%. The proposed approach significantly outperforms the previous methods, indicating its superiority in accurately recognizing human actions in videos. Furthermore, the proposed approach achieved an average accuracy of 99.49%. This result indicates that the proposed approach, which combines the LRCN model with an attention mechanism, outperforms all the other methods in terms of accuracy on the UCF11 dataset. We selected certain hyperparameters for our model as shown in Table 1 and maintained their consistency across all experiments. While these hyperparameters were set as constants, we found through multiple repeated experiments that they had minimal impact on the overall performance of the models. Specifically, we observed only slight changes in performance when different hyperparameter configurations were tested. The chosen optimizer for our models was rmsprop, and we utilized a batch size of 4 during training. Additionally, we trained the models for

Table 2 Performance evaluation of human action recognition methods

Authors	Network architecture/methods	Accuracy (%)
Dai et al.	Two-stream LSTM	96.90
Meng et al.	QST-CNN-LSTM	89.70
Gammulle et al.	CNN-LSTM	89.20
Patel et al.	Features'fusion	89.43
Gharaee et al.	(SOMs), supervised neural networks, attention mechanism	89.50
Pan et al.	Gaussian mixture models, KNN	89.24
Muhammed et al.	BiLSTM DCNN	98.30
Proposed approach	LRCN model attention mechanism	99.49

Note Comparison of the proposed approach with deep learning methods using the benchmark UCF11

100 epochs and employed early stopping techniques to determine the optimal stopping point. These hyperparameters, along with others, were carefully considered to ensure stable and reliable training processes for our models. In addition Based on the accuracy and validation accuracy curves as well loss and validation loss curves of our proposed method displayed in Fig. 4a and b, it is clear that we have constructed a model that achieves high accuracy possible for Human action recognition (as shown in Fig. 2).

The visualization of the feature maps displayed in Fig. 5 obtained from the proposed Human Action Recognition (HAR) network provides valuable insights into the effectiveness of the model in extracting discriminative features. The Long-term Recurrent Convolutional Network (LRCN) architecture employed in the proposed hybrid network demonstrates its capability in extracting both spatial and temporal information from input video sequences. The feature maps exhibit clear patterns and activations that correspond to various action-related attributes, highlighting the network's ability to learn and represent complex visual cues. These visualizations serve as evidence of the proposed LRCN's proficiency in extracting meaningful and informative features, which are crucial for accurate Human action recognition. Figure 6 displayed a heat map during archery actions, revealing how the network allocates attention to various parts of the input frames of the video. In this visualization, different colors signify different levels of attention or activity intensity. The color scheme typically ranges from cool colors (like blue or green) representing regions of lower attention or activity. In the context of archery, they might correspond to less critical stages or elements of the action, where the network's focus is less pronounced, to warm colors (like yellow or red) signifies the highest attention and activity levels. These regions on the heat map correspond to the most crucial aspects of the archery action that the network is prioritizing for accurate recognition and classification. For archery, this could include moments like drawing the bowstring, aiming, and releasing the arrow.

Fig. 4 Accuracy and loss curves

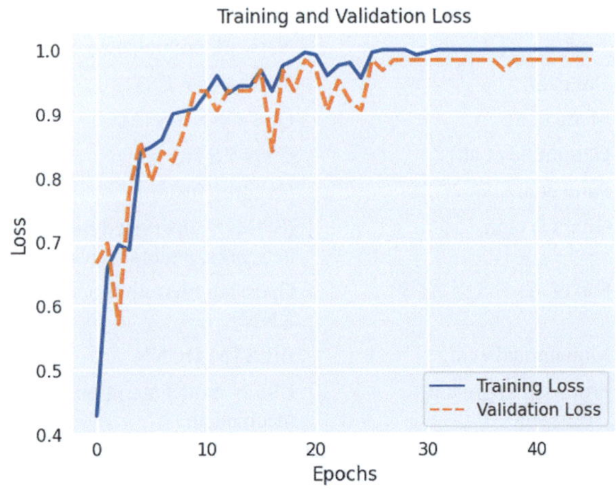

(a) Accuracy and Validation accuracy curves.

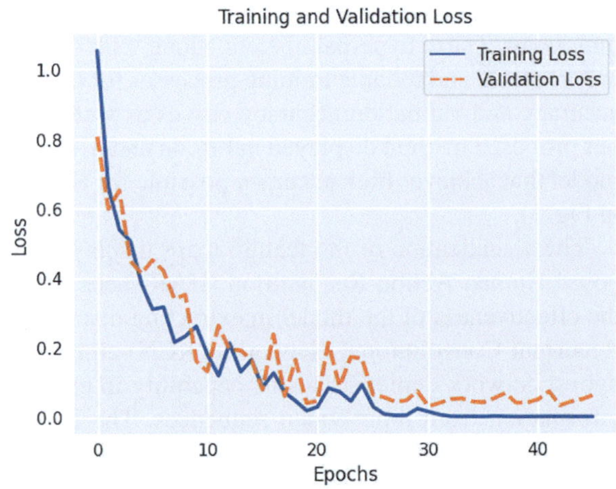

(b) Accuracy and Validation accuracy curves.

5 Conclusion

The proposed approach highlights the paramount importance of extracting spatio-temporal features for accurate action recognition in video, as these features are crucial for numerous applications across various domains. The ability to recognize and understand human activities from video sequences has widespread implications, including video surveillance, sports analysis, human-computer interaction, and more. In our research, we have successfully addressed this challenge by employing a comprehensive approach that combines a time-distributed wrapper and attention-based mechanisms. By incorporating a time-distributed wrapper, our model effectively

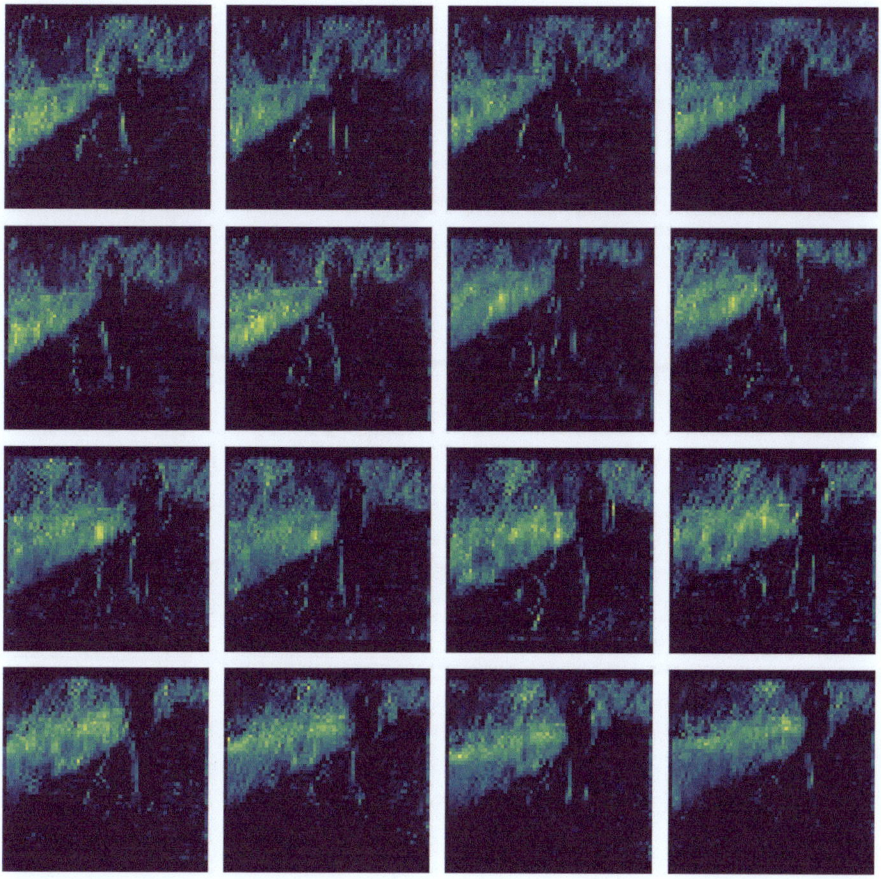

Fig. 5 Visualisation of the feature maps generated by convolution layers for walking with dog activity of the proposed HAR network

extract temporal dynamics by extending the capabilities of convolutional neural networks (CNNs) to process sequences of frames. This enables the network to learn and represent the evolution of actions over time, resulting in enhanced recognition performance. Additionally, the integration of attention-based mechanisms further enhances the model's discriminative power by focusing on the most informative regions and frames within the video. This attention mechanism enables the network to selectively weigh and emphasize relevant spatio-temporal features, leading to improved action recognition accuracy.

Data Availability Statement The datasets generated during and/or analyzed during the current study are available from the corresponding author on reasonable request.

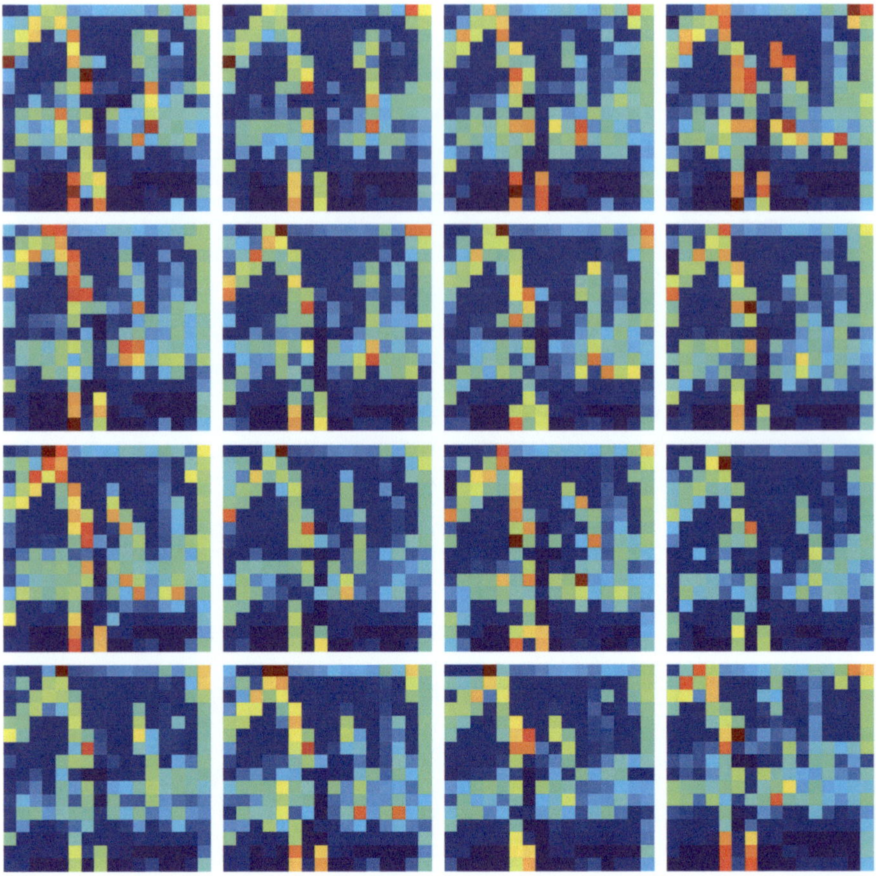

Fig. 6 Visualizing attention heat maps of the proposed HAR network during archery action

Declarations

Ethical Approval This article does not contain any studies with human participants or animals performed by any of the authors.

Funding This research did not receive any specific grant from funding agencies in the public, commercial, or not-for-profit sectors.

Conflicts of Interest All authors declare that they have no conflict of interest.

References

1. Sun Z et al (2023) Human action recognition from various data modalities. IEEE Trans Pattern Anal Mach Intell 45(3):3200–3225
2. Host K et al (2022) An overview of human action recognition in sports based on computer vision. Heliyon 8(6):e09633
3. Bibbo BL et al (2022) An overview of indoor localization system for human activity recognition (HAR) in healthcare. Sensors 22(21):8119
4. Donahue J et al (2015) Long-term recurrent convolutional networks for visual recognition and description. In: Proceedings of the IEEE conference on computer vision and pattern recognition (CVPR), pp 2625–2634
5. Dai C et al (2020) Human action recognition using two-stream attention based LSTM networks. Appl Soft Comput 86:105820
6. Meng B et al (2018) Human action recognition based on quaternion spatial-temporal convolutional neural network and LSTM in RGB videos. Multimed Tools Appl 77(20):26901–26918
7. Gammulle H et al (2017) Two stream LSTM: a deep fusion framework for human action recognition. In: IEEE winter conference on applications of computer vision (WACV). IEEE
8. Patel CI et al (2018) Human action recognition using fusion of features for unconstrained video sequences. Comput & Electr Eng 70:284–301
9. Gharaee Z et al (2017) First and second order dynamics in a hierarchical SOM system for action recognition. Appl Soft Comput 59:574–585
10. Pan Z et al (2020) Robust basketball sports recognition by leveraging motion block estimation. Signal Process Image Commun 115784
11. Muhammad K et al (2021) Human action recognition using attention based LSTM network with dilated CNN features. Futur Gener Comput Syst 125:820–830
12. Luo H et al (2022) Dense semantics-assisted networks for video action recognition. IEEE Trans Circuits Syst Video Technol 32(5):3073–3084
13. Liu J et al (2009) Learning semantic visual vocabularies using diffusion distance. In: IEEE international conference on computer vision and pattern recognition (CVPR)
14. Kamel B et al (2019) Deep convolutional neural networks for human action recognition using depth maps and postures. IEEE Trans Syst Man Cybern Syst 49(9):1806–1819. https://doi.org/10.1109/TSMC.2018.2850149
15. Jaouedi N et al (2016) Human action recognition to human behavior analysis. In: 7th international conference on sciences of electronics, technologies of information and telecommunications (SETIT), pp 263–266. https://doi.org/10.1109/SETIT.2016.7939877
16. Xiao Q et al (2017) Human action recognition using autoencoder. In: 3rd IEEE international conference on computer and communications (ICCC), pp 1672–1675. https://doi.org/10.1109/CompComm.2017.8322824
17. Ji Y et al (2020) A survey of human action analysis in HRI applications. IEEE Trans Circuits Syst Video Technol 30(7):2114–2128. https://doi.org/10.1109/TCSVT.2019.2912988
18. Wang L et al (2020) A comparative review of recent kinect-based action recognition algorithms. IEEE Trans Image Process 29:15–28. https://doi.org/10.1109/TIP.2019.2925285

Stuttering Diagnosis and Classification

Vaibhav Verma, Richa Baranwal, Arjun Singh Rawat, and Jyoti

Abstract Stuttering is a speech disorder characterized by disruptions in the fluency of speech, such as repetitions, prolongations, and blocks. Advances in technology, including machine learning and neuro-imaging, are enhancing diagnostic precision and understanding of the disorder's underlying mechanisms, paving the way for more effective and personalized treatments. This paper provides a thorough review of recent advancements in the field of intelligent processing of stuttered speech. Through an extensive survey of the literature, we explore various approaches ranging from automatic correction and detection to leveraging clinician annotations for improving automatic speech recognition systems. Stuttering diagnosis and classification can be enhanced using machine learning techniques like k-Nearest Neighbors (KNN) and Decision Trees. k-NN classifies speech samples by comparing features such as disfluency frequency and speech rate to labeled instances, identifying patterns indicative of stuttering. Decision Trees, on the other hand, use features like syllable repetitions and silent pauses to create decision rules, providing clear, interpretable classification criteria. These methods improve diagnostic accuracy and enable personalized treatment strategies for stuttering. Confusion metric is used to capture the model's performance and to showcase achieved results with 89.53% of accuracy with decision tree for word repetition and 86.11% for sound repetition. Stuttering diagnosis and classification face several limitations, including variability in speech patterns, which complicates consistent assessment. Diagnostic tools can be subjective and reliant on clinician expertise, potentially leading to inconsistencies.

V. Verma · R. Baranwal · A. S. Rawat (✉) · Jyoti
National Institute of Technology, Delhi, India
e-mail: arjunsinghrawat@nitdelhi.ac.in

V. Verma
e-mail: 232211032@nitdelhi.ac.in

R. Baranwal
e-mail: 232211028@nitdelhi.ac.in

Jyoti
e-mail: 232211012@nitdelhi.ac.in

S. Chand Kumain et al. (eds.), *AI and ML Techniques in Image Processing and Object Detection*, https://doi.org/10.1007/978-981-96-7445-9_12

Keywords Stuttering · Speech disorder · Speech fluency · Repetitions · Prolongations · Machine learning K-NN · Decision tree · Speech rate · Diagnostic accuracy

1 Introduction

A speech fluency condition known as stuttering is typified by difficulty using the typical sounds required for speech and communication. Stuttering is a speech fluency issue characterized by difficulties producing the regular sounds required for speaking and communication. Stuttering does not currently have a treatment. The only progress made has been in speech therapy symptom management [1]. Communication specialists with training in speech pathology can identify stuttering and provide treatment to lessen or avoid fluency issues. While almost 75% of children who stammer between the ages of two and six recover from their stutter, stuttering can last a lifetime for some people. With a lifetime prevalence of stuttering estimated at 0.72%, there are over 55 million stutterers worldwide, with nearly 80% of them residing in less developed nations in Asia, Latin America, and Africa [2]. Managing stuttering in developing nations presents a significant challenge because these regions lack speech-language pathologists, with India having a shortage of these professionals when it comes to treating stuttering. In developing nations, there is a dearth of clinical resources for medical needs, so it is necessary to make the most of what is already available for a speech pathologist to assist more stutterers [3]. Recent developments in computer speed, machine learning, and natural language processing may make it easier for speech-language pathologists to identify and monitor the development of stutterers through auto-mated stuttering identification systems (ASIS). Blocks, prolongations, word and sound repetitions, and interjections are some of the ways that stuttering disorders manifest. By impairing vital communication abilities and negatively impacting social interactions in academic and professional contexts, this illness can significantly lower a person's quality of life. Stuttering may exacerbate social anxiety as people age, which would further impair their general wellbeing [4]. Further-more, stuttering has been linked to lower scores on the Medical Outcomes Study Short Form-36 (SF-36) quality of life questionnaire in the areas of vitality, social functioning, emotional functioning, and mental health.

Decision trees and k-Nearest Neighbors (k-NN) are two machine learning algorithms that can improve the diagnosis and classification of stuttering. By comparing characteristics like speech rate and disfluency frequency to instances that have been categorized, k-NN classifies speech samples and finds patterns that suggest stuttering. Conversely, decision trees provide unambiguous, comprehensible classification criteria by employing characteristics such as syllable repetitions and silent pauses to generate decision rules. These techniques enhance the precision of the diagnosis and allow for individualized stuttering treatment plans. The model's performance is measured using the confusion metric, which also displays the outcomes attained.

2 Literature Review

The literature survey of this "Stutter detection and classification" is based on the below papers.

The paper 'SEP-28 K: A Dataset for Stuttering Event Detection from Podcasts with People WHO Stutter' presents the SEP-28 k dataset [5], outlined for recognizing faltering occasions in discourse, especially from podcasts including people who stammer. It addresses the shortage of commented-on information in this space by giving over 28 k labeled clips, besides explanations for 4 k clips from the Fluency-Bank dataset. The ponder investigates different acoustic models and highlights, counting mel-filterbank vitality, pitch, articulatory highlights, and phoneme probabilities, combined with LSTM and ConvLSTM structures. The comes about appear that consolidating different highlights and utilizing the ConvLSTM show with a CCC misfortune essentially makes strides F1 scores and decreases EER, especially for identifying pieces and word redundancies. Preparing on bigger datasets like SEP-28 k too improves execution, showing the dataset's utility in creating more exact dysfluency location models.

The paper 'A CNN-Based Automated Stuttering Identification System' pro-poses a novel approach employing a Convolutional Neural Arrange (CNN) to naturally recognize and classify stammering disfluencies in discourse [6]. The creators emphasize the effect of stammering on quality of life and the shortage of discourse dialect pathologists, especially in creating nations. They prepared and tried their CNN demonstration utilizing the Sep-28 k dataset, which contains clarified stammering information, and assessed its execution measurements such as precision, exactness, review, and F1 score. The comes about appears that their show outflanked past classifiers, illustrating the adequacy of CNNs [6] in stammering distinguishing proof [7]. The creators moreover highlight the significance of datasets like Sep-28 k in progressing classifier strength and propose regions for assist advancement, such as information increase and investigating distinctive machine learning models. By and large, their investigation exhibits promising headway in robotized faltering location frameworks, with potential suggestions for moving forward to discourse treatment universally [8].

The paper 'Robust Stuttering Detection [9] via Multi-task and Adversarial Learning' investigates novel strategies in distinguishing stammering in discourse. Utilizing multi-task learning (MTL) and antagonistic learning (ADV), the consider points to form strong models for stammering discovery. The MTL system includes mutually learning faltering and metadata data, whereas ADV centers on learning strong and metadata-invariant acoustic representations for faltering. The proposed system is based on time delay neural systems and is assessed utilizing the SEP-28 k stammering dataset, appearing enhancements in different disfluency classes over the pattern. The comes about demonstrates that MTL upgrades location execution for disfluent classes but increments perplexity for familiar tests. In the interim, ADV learns vigorous falter highlights and appears promising results in distinguishing stammering sorts. By and large, the ponder contributes important bits of knowledge into leveraging progressed learning strategies for more successful faltering discovery frameworks.

The paper 'Advancing Stuttering Detection via Data Augmentation, Class-Balanced Loss, and Multi-Contextual Deep Learning' addresses challenges in faltering discovery by proposing novel techniques [10]. It presents MC StutterNet, a time delay-based neural organize, and utilizes information enlargement methods from the MUSAN dataset. The inquiry accomplishes outstanding headways in stammering discovery, with a large-scale F1 score of roughly 91% on the reenacted LibriStutter dataset. Be that as it may, cross-corpora assessments uncover challenges in demonstrating generalization, highlighting the requirement for domain-specific information expansion, and assisting in robotized faltering location [11]. The think about recognizes the complexities of the stammering discovery space, emphasizing continuous endeavors to move forward with clinical ease of use and show explainability.

Existing work in stutter detection and classification faces several limitations. There is a scarcity of large, diverse, and annotated datasets, which affects model robustness and generalization. Models often struggle to generalize to new data, and the optimal combination of acoustic features remains unclear, impacting performance consistency. The complexity of stuttering phenomena, with various types such as repetitions and blocks, poses challenges, as models may perform well for some types but not others. Adversarial learning techniques, while promising, add complexity to training and require careful balancing. Many models lack clinical usability and explainability, crucial for therapeutic adoption. Additionally, the computational demands of sophisticated models hinder real-time or on-device deployment, particularly in resource-limited settings. Existing data augmentation methods may not fully capture the variability of natural speech disfluencies, high-lighting the need for more effective strategies.

3 Proposed Model

In our proposed work, we have used a decision tree for word repetition and sound repetition and KNN for prolongation. Figure 1 shows the block diagram of stutter speech detection. It represents a systematic approach to processing and analyzing stuttered speech, from raw audio input to the identification and classification of stuttering events.

Figure 2 shows the block diagram of the Mel-Frequency Cepstral Coefficients (MFCC). In speech and audio processing, MFCC has often utilized features, especially for tasks like speaker identification and speech recognition [12].

1. Frame Segmentation: The audio signal is split into brief segments, typically lasting between 20 and 30 ms, with overlaps to better capture time-related details.

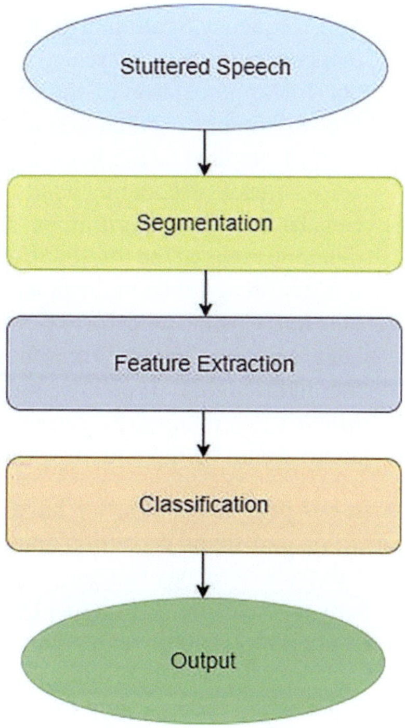

Fig. 1 Block diagram for stutter speech detection

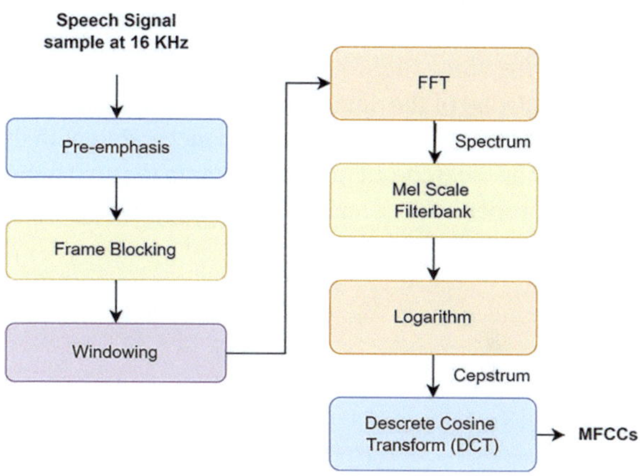

Fig. 2 MFCC block diagram

2. Windowing: A window function (such as a Hamming window) is applied to each segment to minimize spectral leakage that can occur during the analysis of short-duration signals.
3. Discrete Fourier Transform (DFT): Each windowed segment is trans-formed from the time domain into the frequency domain using the DFT, thereby revealing the signal's spectral content.

4. Mel-frequency Scaling: The frequency spectrum obtained from the DFT is then mapped onto the mel scale, which aligns more closely with the human ear's non-linear perception of sound frequencies.
5. Mel-filterbank: A collection of triangular filters, arranged according to the mel scale, is applied to the transformed spectrum to capture the energy distribution across various frequency bands.
6. Logarithm: The logarithm of the filterbank outputs is taken to compress the dynamic range of the filterbank energies. This step helps in dealing with variations in signal magnitude and enhances the sensitivity to lower energy regions.
7. Discrete Cosine Transform (DCT): Finally, the DCT is applied to the log filterbank energies. The DCT coefficients obtained represent the cepstral features of the audio signal. Typically, only the lower-order DCT coefficients, known as MFCCs, are retained as they capture the essential spectral charac-teristics of the audio signal while reducing dimensionality.

In our proposed work, we have used a decision tree shown in Fig. 3 for word repetition and sound repetition and KNN shown in Fig. 4 for prolongation.

A decision tree may be a flowchart with hubs and bolts. Each record of the dataset streams through the flowchart [13]. Each record of the information begins at the root hub, which is on best, and voyages through inside hubs to the conclusion in a last leaf hub. In each hub other than the leaf hubs, a choice is made around where the information record ought to go to another [14]. All information records start off within the root hub and after that travel to either the cleared-out or right inner hub specifically underneath, based on whether the foremost critical include is less than or more prominent than a conditional expression, such as chroma cq ≤ 0.6. Chroma cq is one of the show highlights. By partitioning the information lower or higher than 0.6, the entropy of the dataset is maximized. That is, after this division, the following two inside hubs are as diverse as is conceivable with the set of show highlights being utilized. The cleared-out hub will contain the next concentration of the target variable than the proper hub, or bad habit versa. At that point usually assist subdivided at the

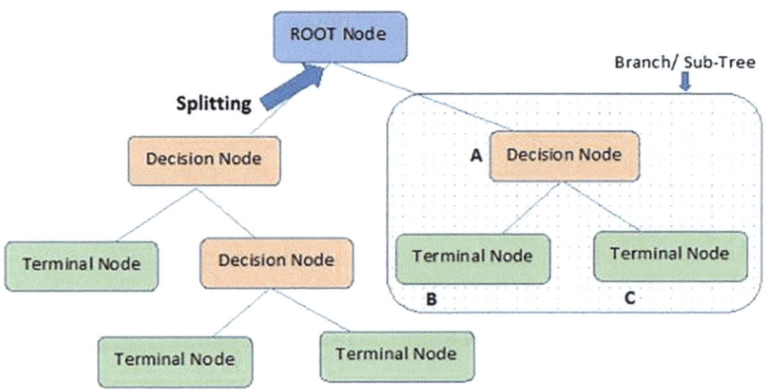

Fig. 3 Decision tree

Fig. 4 K-nearest neighbour (k-NN) algorithm

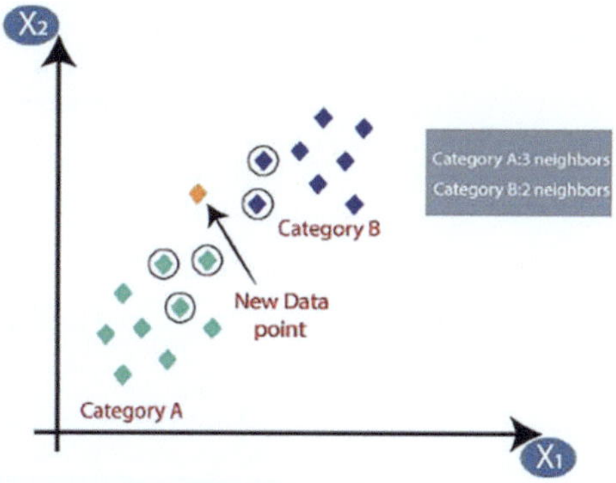

another push [15]. This handle stops when the number of tests in a leaf gets as well. The Python code sets modeling parameters just like the least number of tests in any leaf, or the number of lines the tree ought to have. It too takes within the title of the target variable and the names of all the modeling highlights.

K-nearest neighbours (KNN) may be a clear however compelling calculation utilized in machine learning for classification and relapse errands. The concept behind KNN is straightforward: when foreseeing the course of a modern information point, the calculation looks at the K closest information focuses within the preparing set based on a remove metric (frequently Euclidean remove). For classification, the larger part course among the K neighbors is allotted to the modern information point, whereas for relapse, the calculation calculates the normal (or weighted normal) of the target values of the K neighbors.

KNN is non-parametric and instance-based, meaning it doesn't make suspicions around the basic information dispersion and stores the complete preparing dataset for the forecast. Be that as it may, choosing a suitable esteem for K is pivotal because it impacts the bias-variance trade-off of the show.

4 Dataset

The information utilized in this study is from the Sep-28 k and FluencyBank datasets. The Sep-28 k dataset [16] contains 28,000 three-second sound clips from podcasts made by stammering people, both in male and female voice. The clips were chosen at irregular from these podcasts [5]. An extra 4,000 clips from other podcasts were created comparably for the FluencyBank dataset [17]. This collectively brings the overall number of three-second sound clips to 32,000. Three discourse master judges tuned in to all 32,000 sound clips and voted whether they thought diverse sorts of faltering were show, or no stammering at all. Types of stammering are shown in Table 1.

Table 1 Dataset description

Disfluency	Definition
Word repetition	Repetition of word
Sound repetition	Phoneme is repeated
Prolongation	Extended sounds

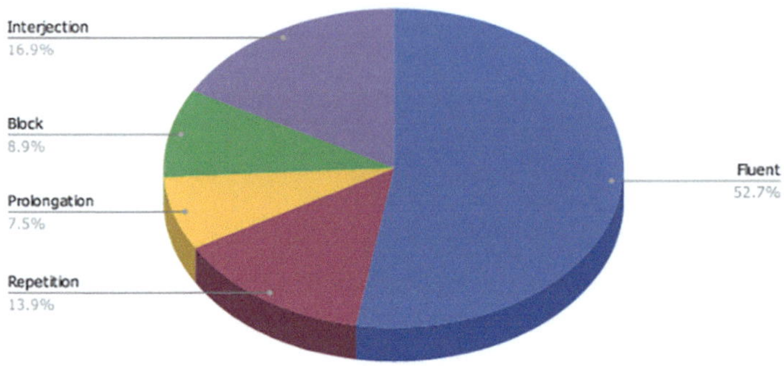

Fig. 5 Sep-28 k dataset

Each of these types of faltering for each of the 32,000 podcasts encompasses a name which may be a number from zero to three, speaking to the number of judges who voted for that sort of faltering for that sound clip. For illustration, the 10th clip of the to begin with a scene from the "HeStutters" podcast and encompasses a two beneath square stammer. This suggests that two of the three judges accepted that piece stammering happened amid this three-second sound clip. A few clips had more than one stammering sort, and a few clips had no faltering at all. The Sep-28 k (Stuttered Events Podcasts-28,000) dataset is published by Apple in 2021. Figure 5 shows the dataset where 52.7% is the fluent dataset and the remaining 47.3% dataset is about interjection, block, prolongation, and repetition. These all are types of stuttering.

The Sep-28 k dataset's composition is displayed in Fig. 5. The dataset's various categories are distributed as follows:

- Fluent (52.7%): The largest category, representing over half of the dataset.
- Interjection (16.9%): The second largest category.
- Repetition (13.9%): The third largest category.
- Block (8.9%): A smaller category.
- Prolongation (7.5%): The smallest category in the dataset.

Varieties of speech disfluencies, including prolongations, sound repeats, and word repetitions. The Sep28k dataset for stutter identification, which hasn't been utilized or examined frequently in relevant work. Our model uses a decision tree and KNN to classify three-second speech fragments with the appropriate analysis of disfluency or not, in contrast to other models that have been created on the Sep28k dataset.

Figure 6 shows the number of audio clips from various shows. "WomenWhoStutter" and "StutterTalk" contribute the most clips, with over 8000 and 6000 respectively.

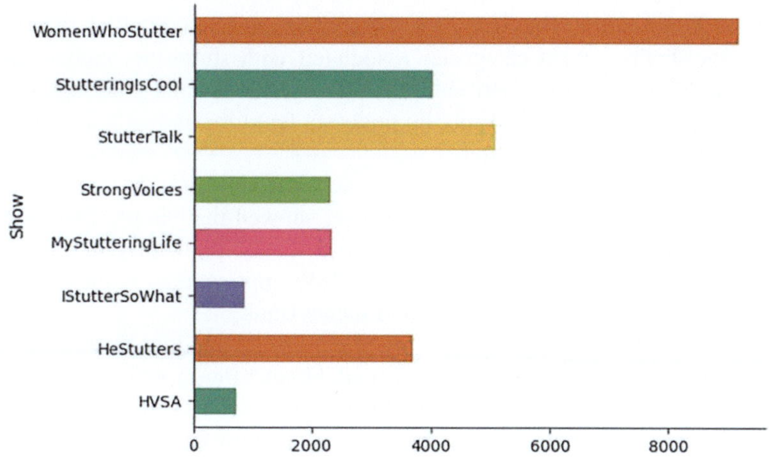

Fig. 6 Sep28k dataset

Other significant contributors include "HeStutters" and "StutteringIsCool," while "HVSA" and "IStutterSoWhat" provide the fewest clips. This indicates a diverse range of sources in the dataset.

The task is to categorize disfluency into one of five categories (word repetition, sound repetition, prolongation, block, and interjection). The dataset consists of 32 k examples which is covered in podcasts. This dataset is publicly available on Kaggle and Fluencybank [18] official site.

5 Experimental Results

SEP-28 k dataset is used for detecting stuttering events in speech. Experimentation showed a significant performance boost with SEP-28 k compared to previous datasets. As shown in Table 2, the Decision Tree model is more effective for detecting words with an accuracy of 89.53% and sound repetitions with an accuracy of 86.11%, while the K-Nearest Neighbour model is used for detecting prolongations with lower accuracy.

Table 2 Accuracy of each dis-fluency

Dis-fluency	Model	Accuracy
Word repetition	Decision tree	89.53
Sound repetition	Decision tree	86.11
Prolongation	K-nearest neighbour	66.83

1. Dataset Description: The SEP-28 k dataset has more than 28,000 clips labeled with five distinct event categories associated with stuttering, namely prolongations, blocks, word repetitions, sound repetitions, and interjections. The audio clips are sourced from public podcasts featuring individuals who stutter interviewing others who stutter.
2. Performance Improvement: Comparing acoustic models using SEP-28 k and the publicly available FluencyBank dataset showed that simply increasing the amount of training data improves relative detection performance. Com-paring the results to earlier datasets, there was a 28% improvement in the F1 score on SEP-28 k and a 24% improvement on FluencyBank.
3. Annotation Release: It emphasizes the release of annotations from over 32,000 clips across both datasets, SEP-28 k and FluencyBank, which will be made publicly available. This makes it easier to conduct additional study and advancement in the field of speech dysfluency identification.
4. Evaluation Metrics: When evaluating a predictive model's performance for machine learning and classification tasks, a confusion matrix is a useful tool. When handling binary or multi-class classification problems, it is especially helpful. The model's predictions are tabulated and compared to the dataset's actual labels in the confusion matrix. Let's examine the terms that are frequently seen in a confusion matrix.

    (a) Accuracy: This metric quantifies the ratio of correctly classified instances to the total number of cases.

    $$\text{Formula} : \text{Accuracy} = (\text{TP} + \text{TN})/(\text{TP} + \text{TN} + \text{FP} + \text{FN})$$

    (b) Precision: Also known as positive predictive value, precision measures the proportion of true positive predictions among all cases predicted as positive. Formula: Precision $=$ TP/(TP $+$ FP)
    (c) Recall (Sensitivity): This indicator, also called the true positive rate, calculates the fraction of actual positive cases that are correctly identified. Formula: Recall $=$ TP/(TP $+$ FN)
    (d) Specificity: Often referred to as the true negative rate, specificity deter-mines the proportion of actual negatives that are accurately predicted. Formula: Specificity $=$ TN/(TN $+$ FP)
    (e) F1 Score: The F1 Score is the harmonic mean of precision and recall, offering a balanced measure of their combined performance.

    $$\text{Formula} : \text{F1 Score} = 2 \cdot \frac{\text{Precision} \times \text{Recall}}{\text{Precision} + \text{Recall}}$$

In Fig. 7 the color intensity in the matrix represents the count of predictions, with darker colors indicating higher counts. The performance metrics derived from this confusion matrix help evaluate the effectiveness of the classification model.

Figure 8 shows the performance of a classification model.

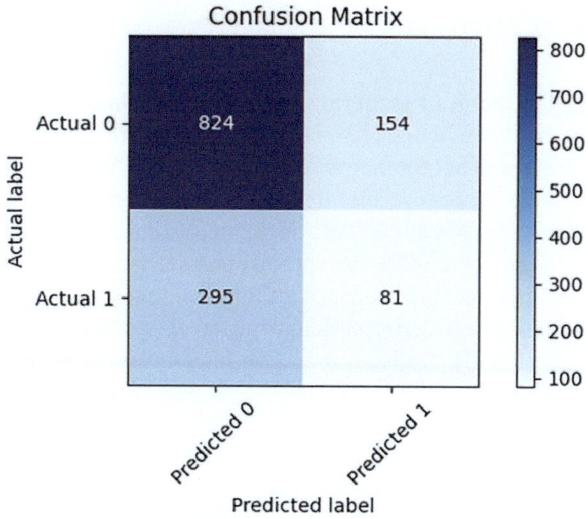

Fig. 7 Confusion matrix

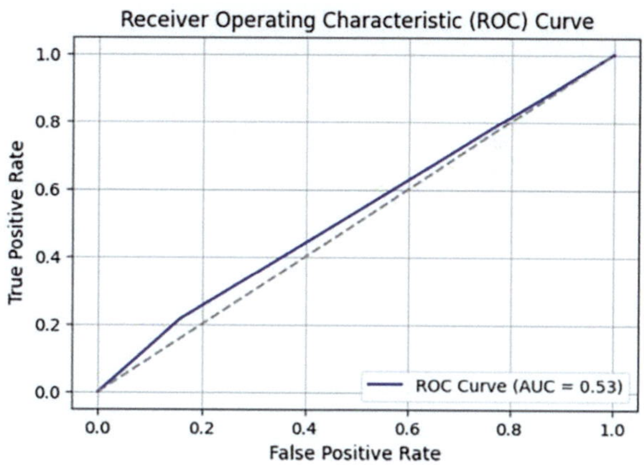

Fig. 8 ROC (receiver operating characteristic) curve

- True Positive Rate vs. False Positive Rate: The curve plots these rates at different thresholds.
- AUC (Area Under the Curve) = 0.53: Indicates the model is only slightly better than random guessing (0.5).

This ROC curve helps assess the model's ability to distinguish between classes.

6 Conclusion

In the realm of stuttering detection and classification, both decision trees and K-nearest neighbors (KNN) offer distinct advantages and considerations. Decision trees stand out for their interpretability, providing valuable insights into the decision-making process by highlighting which features contribute most significantly to stuttering classification. This transparency can aid researchers and clinicians in understanding the underlying speech characteristics associated with stuttering. Moreover, decision trees are adept at capturing nonlinear relationships between features and stuttering, which can be particularly relevant given the complex nature of speech patterns. Additionally, decision trees exhibit robustness to irrelevant features, automatically selecting the most discriminative ones for split-ting nodes, thereby simplifying feature selection.

On the other hand, K-nearest neighbours (KNN) offer flexibility in handling stuttering detection tasks, especially when dealing with intricate, nonlinear relationships between features and stuttering. Since KNN doesn't make assumptions about the distribution of the underlying data, it can be used in situations where it's unclear how speech features and stuttering are related. Additionally, KNN's capacity to take into account local patterns in the feature space may be useful for spotting minute changes or specific patterns in speech signals that might be connected to stuttering episodes. Additionally, the simplicity of KNN's implementation makes it accessible and efficient, particularly for smaller datasets or situations where computational resources are limited.

The decision between decision trees and KNN for stuttering detection and classification in real-world applications depends on a number of variables, including as the dataset's size and complexity, available computing power, and the required degree of interpretability. Researchers and practitioners may benefit from experimenting with both algorithms on their specific dataset, possibly exploring ensemble methods like Random Forests for decision trees or considering more advanced algorithms like Support Vector Machines (SVMs) or Neural Networks for further enhancement in classification performance. In order to choose the best method or algorithms, these criteria must be carefully considered in relation to the particular needs and objectives of the stuttering detection task.

7 Future Scope

Future developments in the field of stuttering detection and classification are anticipated to be significantly accelerated by emerging technology and research projects. Advanced machine learning methods, such as convolutional neural networks (CNNs) and recurrent neural networks (RNNs), can more reliably and ac-accurately detect stuttering speech patterns in fluent speech. By combining various modalities—such as audio, video, and physiological signals—we can improve our comprehension of stuttering behaviours and the precision of our detection systems. With the help of real-time detection systems that can be installed on portable electronics like smartphones, stutterers can now manage their speech in a variety of social and professional settings by receiving prompt feedback. These systems' longitudinal monitoring can offer insightful information about the development of stuttering, guiding the development of tailored intervention plans. While ethical considerations and user entered design principles ensure the responsible and inclusive development of these technologies, large-scale data analytics initiatives may reveal new insights into the complexities of stuttering. Eventually, adding stuttering detection tools to teletherapy platforms can increase the number of options for remote assessment and treatment, leading to better clinical outcomes and an improvement in the quality of life for stutterers.

Authors' Contributions All authors contributed to the study conception and design. Material preparation, data collection, and analysis were performed by Vaibhav Verma, Richa Baranwal, and Jyoti. The first draft of the manuscript was written by Arjun Singh Rawat, and all authors commented on previous versions of the manuscript. All authors read and approved the final manuscript.

Funding This research received no specific grant from any funding agency in the public, commercial, or not-for-profit sectors.

Data Availability Data and code will be made available upon reasonable request from the authors.

Conflict of Interest The authors declare that they have no conflict of interest.
Compliance with Ethical Standards: Not applicable.

References

1. Wheeler K (2020) For people who stutter, the convenience of voice assistant technology remains out of reach. USA Today (Online)
2. Mahesha P, Vinod DS (2016) Gaussian mixture model based classification of stuttering dysfluencies. J Intell Syst 25(3):387–399
3. Craig A, Hancock K, Tran Y, Craig M, Peters K (2002) Epidemiology of stuttering in the community across the entire life span
4. Craig A, Blumgart E, Tran Y (2009) The impact of stuttering on the quality of life in adults who stutter. J Fluen Disord 34(2):61–71
5. Lea C, Mitra V, Joshi A, Kajarekar S, Bigham JP (2021) Sep-28k: a dataset for stuttering event detection from podcasts with people who stutter. In: ICASSP 2021–2021 IEEE international conference on acoustics, speech and signal processing (ICASSP). IEEE, pp 6798–6802
6. Prabhu Y, Seliya N (2022) A CNN-based automated stuttering identification sys-tem. In: 2022 21st IEEE international conference on machine learning and applications (ICMLA). IEEE, pp 1601–1605
7. Nassif AB, Shahin I, Attili I, Azzeh M, Shaalan K (2019) Speech recognition using deep neural networks: a systematic review. IEEE Access 7:19143–19165
8. Liu X (2018) Deep convolutional and LSTM neural networks for acoustic modelling in automatic speech recognition
9. Sheikh SA, Sahidullah Md, Hirsch F, Ouni S (2022) Robust stuttering detection via multi-task and adversarial learning. In: 2022 30th European signal processing conference (EUSIPCO). IEEE, pp 190–194

10. Sheikh SA, Sahidullah Md, Hirsch F, Ouni S (2023) Advancing stuttering detection via data augmentation, class-balanced loss and multi-contextual deep learning. IEEE J Biomed Health Inform
11. Brewer RN, Findlater L, Kaye JJ, Lasecki W, Munteanu C, Weber A (2018) Accessible voice interfaces. In: Companion of the 2018 ACM conference on computer supported cooperative work and social computing, pp 441–446
12. Chopra M, Khieu K, Liu T (2020) Classification and recognition of stuttered speech. Manu Chopra.pdf
13. Quinlan JR (1986) Induction of decision trees. Mach Learn 1:81–106
14. Breiman L (2001) Random forests. Mach Learn 45:5–32
15. Alharbi S, Hasan M, Simons AJH, Brumfitt S, Green P (2018) A lightly supervised approach to detect stuttering in children's speech. In: Proceedings of interspeech 2018. ISCA, pp 3433–3437
16. Sep28k dataset. Accessed 29 Apr 2024
17. FluencyBank. Accessed 29 Apr 2024
18. Ratner NB, MacWhinney B (2018) Fluency bank: a new resource for fluency research and practice. J Fluen Disord 56:69–80

MIX
Papier aus verantwortungsvollen Quellen
Paper from responsible sources
FSC® C105338

If you have any concerns about our products,
you can contact us on
ProductSafety@springernature.com

In case Publisher is established outside the EU,
the EU authorized representative is:
Springer Nature Customer Service Center GmbH
Europaplatz 3, 69115 Heidelberg, Germany

Printed by Libri Plureos GmbH
in Hamburg, Germany